BASIC CONCEPTS OF MATHEMATICS AND LOGIC

Michael C. Gemignani

DOVER PUBLICATIONS, INC.
Mineola, New York

Bibliographical Note

This Dover edition, first published in 2004, is an unabridged republication of the edition published by Addison-Wesley Publishing Co., Reading, Massachusetts, 1968.

Library of Congress Cataloging-in-Publication Data

Gemignani, Michael C.
    Basic concepts of mathematics and logic / Michael C. Gemignani.
        p. cm.
    Originally published: Reading, Mass. : Addison-Wesley Pub. Co., 1968.
    (Addison-Wesley series in introductory mathematics)
    Includes index.
    ISBN 0-486-43506-7 (pbk.)
    1. Mathematics. 2. Logic, Symbolic and mathematical. I. Title.

QA39.3.G46 2004
510—dc22

2003070108

Manufactured in the United States of America
Dover Publications, Inc., 31 East 2nd Street, Mineola, N.Y. 11501

Preface

A college education has two fundamental goals: vocational preparation and intellectual stimulation. Mathematics is an essential tool for many vocations but it is also a very fine subject of study for stimulating creative thought. It demands original and rational thought processes, many of which can be applied to other topics of study.

This text is intended as a first look at mathematics at the college level; the emphasis is strongly on logic and the theory of sets. It is hoped that the student who does not plan to take further courses in mathematics will acquire from this text an appreciation of what mathematics is and of the thought processes essential to mathematics; such an appreciation will enrich his intellectual resources and aid him in meeting other challenges to his reasoning abilities. For those who wish to continue in mathematics, it is hoped that this book will furnish a solid foundation in certain central concepts of mathematics: set theory, logic, counting, numbers, functions, ordering, probabilities, and other building-blocks of higher mathematics.

The first view of a discipline should reveal a few well-chosen topics rather than expose a superficial view of too many of its elements. Hence this text does not cover as many topics as many comparable introductions to college-level mathematics. All the topics discussed in this text are important ones and they are covered in greater depth, with more motivation, and with greater rigor than is usually the case at the introductory level.

M. C. Gemignani

Buffalo, N.Y.
January, 1968

To Stephen and Susan

Contents

chapter one

Introduction

1.1 ABSTRACTION AND MATHEMATICS

The philosopher Aristotle believed that the power to abstract was one of the essential characteristics of human beings, a power indeed which distinguished men from the lower forms of living creatures. It is certainly true that the power to abstract, that is, the ability to draw the essence from what we see or experience and to know that essence apart from any individual object which possesses it, is one of our most important talents. It is, however, such a common talent that we rarely even think about it; we merely abstract as we would breathe or eat.

Consider our dilemma if we could not abstract. We would have to have a different name for absolutely everything we encounter. If we were incapable of realizing what makes a table a table, we would have to call each table we see by a different name in order to communicate to some-one else what object we were referring to. To say "the table in the kitchen" would be meaningless because we would not know what an object had to be in order to be a table. Each time an infant encounters a table, his parents say to him, "This is a table," and it is a proud moment when the child sees a table he never saw before and calls it a table. He has exercised his human powers of abstraction at an early age.

We might say that if Aristotle was right, then the dictionary, which gives the essence of things, is the most human of all books. But, likewise, we might also be able to say that mathematicians are the most human of all men, for the history of mathematics has been a history of abstraction from the first time that primitive man began to count, to symbolize objects by the abstract concept of their quantity, up to the present day in which mathematics has reached a highly complex state.

Most mathematics developed originally as a response to a practical problem, or as an attempt to characterize some observable phenomenon. In order to hold a discussion about a horse, we must know what a horse is; furthermore, the person to whom we are speaking must have the same idea of "horse" that we have. Thus, as we study the phenomena of our universe, it is natural that we try to set down basic or defining properties of these phenomena so that, first, we are agreed as to what is under discussion, and second, we have a starting point from which we may proceed to draw logical conclusions.

When Euclid looked at the world about him, he observed a certain order from which he tried to abstract the basic properties. When Newton studied planetary motion, he tried to set down the fundamental rules which determined this motion. From their initial abstractions, Euclid and Newton went on to draw far-reaching and important conclusions.

Abstraction enables us to consider, recognize, and draw conclusions about a great many individual objects simultaneously because it makes us aware of what properties these objects have, whether the objects be horses, the geometry of the world we live in, or the motions of the planets.

We can, however, have various degrees of abstraction. If we consider only one particular dog, we are not abstracting at all. If we study "dogness," that is, what makes a dog a dog, we surely include a larger group of objects, namely, all dogs. But a dog is a mammal, a mammal is an animal, an animal is a living thing, and with each transition we move to a larger class of objects which includes everything before it; hence we are getting more and more abstract.

If we consider "dogness" and draw conclusions from it, then these conclusions will apply to any individual dog, whereas the conclusions drawn from studying one particular dog may apply to very few dogs. Whatever we learn about living things will apply to dogs but, unfortunately, may tell us very little about dogs as dogs since the generality of "living things" is so great.

It should be realized though that this process of abstracting has a great unifying effect, for, although we learn little about individual dogs by studying the essential nature of living things, what we do learn from such a study is of considerable importance because of its wide applicability. The more general the class of objects we study, the less we learn about individual objects, but the results of our study are of more universal interest; moreover, by including many subcategories in our study, we unify to a large extent the separate studies of each subcategory and gather them all under one roof. The development of mathematics has usually followed this pattern of abstraction, proceeding from the study of the concrete to the study of the more and more general.

1.2 MATHEMATICS AND "REALITY"

The mathematician creates mathematics in his mind and mathematics essentially has its existence there. This may at first sound contradictory to what has been said regarding the abstractive quality of mathematics, but it is not contradictory for the following reasons: First, while mathematics may receive its initial impetus from the "real" world, once a mathematical system so inspired has been defined, it is something existing in its own right, and with a value all its own quite apart from what inspired it. The original abstraction may also give rise to other definitions, each further removed from the initial external inspiration. Second, since abstraction includes the power to consider properties apart from any individual objects in which these properties are found, then definitions can be made up and studied even if they have no apparent realization outside our minds.

Of all scholars, the mathematician is most free to do as he pleases for, as we have pointed out, the mathematician creates what he studies. The physicist, the chemist, the economist, all profess to study "real" things, and their theories rise or fall on the basis of how well they can explain and predict observable phenomena. The mathematician does not depend on the real world, at least so far as his mathematics is concerned.

Nevertheless, all scholars, including mathematicians, approach their studies either with a set of basic assumptions, or with the hope of formulating basic laws and assumptions. Even the most experimental of scientists hopes that the data that he collects will enable him either to find a better set of laws for predicting physical phenomena or to confirm some already existing theory. Indeed, science without unifying laws or assumptions is merely a hodgepodge of data without meaning or value. The numerous data compiled by the astronomer Tycho Brahe and others were useless until Kepler found order in them by means of his famous three principles of planetary motion. But Kepler's laws were in turn found to be simple consequences of the more general and far-reaching theory of motion produced by Newton.

Thus science, and in fact all knowledge, needs a set of fundamental principles or assumptions on which to rest. In mathematics, too, the theories are derived from fundamental assumptions. For example, the reader is probably aware that all plane geometry flows from a collection of axioms.

Yet mathematics differs from other disciplines in the manner in which its basic assumptions are both obtained and used. For example, the underlying laws of a physical theory are generally derived from empirical observations and are intended to have predictive value for the real world. The principal claim to fame of any axioms will be the success which they have in explaining and predicting those phenomena toward which they are directed. Einstein's relativistic view of physics is "better" than Newton's because Einstein's assumptions allow predictions which accord more precisely with observable data than do Newton's assumptions. Furthermore, the assumptions and terms used in the formulation of a physical theory usually refer to concrete objects and situations. We are not really free to interpret the meaning of a theory however we please since we are supposedly referring to objects actually existing in the real world.

The mathematician, however, is free to compose systems of axioms which need not have any relation to the real world. The mathematician can create a set of assumptions and logically develop their consequences without fear that some experiment will one day upset his theory. Moreover, since the mathematician is not obliged to refer to the real world, some of the terms in his axioms may be left undefined, that is, without any one specified meaning. If someone later wishes to apply the mathematician's theory to some concrete situation, he will have to assign to each undefined term a specific meaning that makes the mathematical system fit his case.

Indeed, the fact that mathematical definitions have value all by themselves means that we cannot assign a definite meaning to every term used in setting up the definition. This would in essence be pinning the definition to something external. While a mathematician may make up a definition or axiomatic structure to study some observable phenomenon, after he has made the definition, the phenomenon which inspired it becomes only one example which fits the definition; but something else may also fit the definition when the terms are given new interpretations.

The mathematician thus differs from the physicist, philosopher, and any other scholar who defines things and then draws conclusions from the definitions. Their terms must refer to external realities and hence are defined. Since presumably there can be only one real and true state of things, each word must represent something precise and unambiguous, and we cannot arbitrarily assign interpretations to the terms of their theories.*

However, even the mathematician is bound by certain laws, the laws of correct reasoning, that is, the laws of logic. We may compare his situation to that of an architect who is given unlimited freedom to build any sort of structure he pleases anywhere in the world. Even an architect with such freedom as this is still constrained by the laws of physics: he must build in accordance with the conditions imposed on him by the very framework in which he exists.

Though the mathematician can give free rein to his imagination in formulating axioms to define his mathematical systems, he is nevertheless bound by laws which are "above" mathematics, laws which lie at the very roots of the thought process.

First, the axioms which the mathematician composes must be *consistent*, that is, they must not contradict one another. If the truth of some

---

* The author is aware that these remarks do not apply so much to modern physicists, who essentially are mathematicians, as they do to the classical physicists.

of the axioms belies the truth of another of the axioms, then such a system of axioms would be quite untenable even for a mathematician. For example, suppose a mathematician uses the following three axioms to define a mathematical system:

Axiom 1.   Any two points lie on at most one line.

Axiom 2.   Given any two lines, there are two points which lie on both of them.

Axiom 3.   There are precisely two lines.

Note that the terms "line," "point," and "lie on" are undefined. However, we can see that Axioms 2 and 3 cannot both be satisfied if Axiom 1 is to be satisfied. For Axiom 3 says that there are at least two lines. Because of Axiom 2, there are two points which lie on both of these lines. But Axiom 1 says that two points cannot lie on more than one line. Thus these three axioms are inconsistent; they cannot all be satisfied simultaneously regardless of how "line," "point," and "lie on" are interpreted.

Even if all the axioms used as the basis of a mathematical system are consistent, the conclusions derived from the axioms, the methods of proof and the validity of the chain of conclusions, are all subject to the laws of logic. To violate the laws of logic would be to invalidate whatever conclusions are reached by the specious reasoning. And here we see another difference between the mathematician and certain other scholars.

For example, for the physicist, "truth" is being in accord with empirical data, or, in some cases, achieving some desired result. In the physical or biological sciences it is quite possible to arrive at a very useful result, a truth, quite by accident. The discovery of radioactivity and penicillin are only two of many incidents where chance (and the receptive mind of the experimenter), rather than logical argumentation, intervened with striking and far-reaching consequences. At times the most intuitive arguments or questionable lines of reasoning have led to exceedingly practical and beneficial results. For the mathematician, however, "truth" consists only of logical validity. It cannot be otherwise; for the mathematician is not the least concerned (at least as a mathematician) with whether his invention is a model of "reality" or whether the results that issue from it are practical. The mathematician cares only about what consequences can or cannot be logically drawn from his axioms. In the context of a particular axiomatic system, a statement is true or false only insofar as it can be proved or disproved from the axioms by using logically valid arguments.*

---

* This is not to say that valuable mathematical results cannot come from hunches, or that intuition cannot precede rigor; but, ultimately, a mathematical result must be formulated and proved precisely or, at least, it must be accepted by mathematicians as being rigorously provable.

Are we then to conclude that mathematics has no practical value and the mathematician is out of touch with reality? Much of our previous experience and education tells us otherwise. But might we at least say that some mathematicians are engaged in work which is highly fanciful and which may never bear practical fruit? Perhaps, but we cannot be sure, for what seems fanciful today may be the means of expressing a highly useful theory tomorrow; such, for example, was the case with the so-called "imaginary numbers." Moreover, to the mathematician, mathematics is an art. As in painting, indeed, as in any art, different schools of thought and approach have their day, the avant-garde becomes the obsolete, and the main streams turn into backwaters; at times, theories which at first are ignored become the cornerstone for valuable advances. By constructing axiomatic systems, the mathematician furnishes possible frameworks for the study of concrete objects. It is in this way that mathematics is a practical science.

1.3 AN EXAMPLE

We might wish to construct a mathematical system based on the following axioms:

Axiom 1. Any two blinks are common to at least one sark.

Axiom 2. Any two sarks are common to at least one blink.

Axiom 3. There can be no more than four blinks.

Axiom 4. No sark is common to every blink.

We shall call anything which is interrelated in accordance with these axioms a "flock." Note first that "blink," "sark," and "common to" are left undefined. Whether or not an object under consideration is in fact a flock will depend on whether we can assign meanings to the undefined terms of our axioms so as to make the axioms applicable in the case involved.

blink

3 blinks = 1 sark

Fig. 1.1

First, let us suppose that we have four dogs (see Fig. 1.1). If we call each dog a blink, each trio of dogs a sark, and interpret "is common to" as meaning "share in common," it is easily seen that with this interpretation, our collection of dogs forms a flock. In this case, the four axioms would read:

Axiom 1. *Any two dogs share in common at least one trio of dogs.*

Axiom 2. *Any two trios of dogs share at least one dog in common.*

Axiom 3. *There are no more than four dogs.*

Axiom 4. *No trio of dogs shares in common all four dogs.*

However, if we choose to define blinks as the vertices of some triangle and sarks as pairs of vertices, then we see that the three vertices of a triangle also form a flock.

We may, however, derive conclusions about flocks without ever concerning ourselves with whether there is an honest-to-goodness example of a flock. We can work with the axioms without having a concrete object to which they can be applied.* Statements proved from our axioms will be in terms of "blink," "sark," "common to" and any definitions we choose to make along the way, and what the statements mean in any individual instance must await our assigning appropriate meanings to suit the particular case in question. For example, one theorem that could be proved from our axioms is the following:

Theorem 1.1. *If a flock contains exactly three blinks, then each sark is common to exactly two blinks.*

This theorem indicates that if we want the vertices of a triangle to be definable as a flock and also want each vertex to be a blink and also want "is common to" to retain its usual meaning, then each sark must consist of a pair of vertices.

The definition of a flock is a true case of abstraction insofar as we are considering "flockness" apart from its embodiment as any concrete object. It differs, however, from usual dictionary definitions in that certain terms of the definition are left to be interpreted at the discretion of the reader.

Beginning with the definition of a flock, we can achieve further abstraction by limiting a new definition to only part of the original definition. For example, we could call a *semiflock* anything which satisfies Axioms 1, 3, and 4, but not necessarily Axiom 2. Then every flock is also

---

* In practice, the mathematician would want to know that there is at least one example of a flock so he can be certain the axioms are consistent.

a *semiflock*, but not every semiflock is necessarily a flock (Fig. 1.2.). What is true of
semiflocks, that is, what can be proved from
the axioms defining a semiflock (1, 3, and 4),
is also true for flocks, but it might also be
true of many things which are not flocks; in
particular, since Axiom 2 need not be satisfied by a semiflock, we could have a semiflock in which two sarks are not common
to any blink.

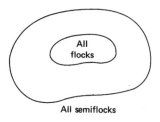

All semiflocks

Fig. 1.2

We might also prove statements about flocks until we arrive at some
property which we feel is worthy of study in its own right. We may then
proceed to study that property even though many things have that
property and are not flocks.*

EXERCISES 1.3

1. Let the vertices of a triangle be a flock, each vertex be a blink, each pair of
   vertices be a sark, and "is common to" have its usual meaning. Express
   Axioms 1 through 4 for this particular case.

2. Find an example of a semiflock which is not a flock. That is, find something
   which, when blink, sark, and "common to" are properly interpreted, satisfies
   Axioms 1, 3 and 4, but not Axiom 2.

3. Decide which of the following satisfy Axioms 1 through 4 with the interpretations of blink and sark as given. "Is common to" is to retain its usual
   meaning in each.
   a) Take the letters A, B, C, and D. Call each of the pairs A, B and C, D a
      blink and call all four letters taken together a sark.
   b) Take four dogs and call each dog a blink; call each pair of dogs a sark.
   c) Take all the letters of the alphabet; call each letter a blink and each trio
      of letters a sark.

4. Prove Theorem 1.1.

5. We have been able to produce a concrete example of a flock. Explain why
   this assures us that the axioms defining a flock are consistent, that is, that
   they do not contradict one another.

6. A collection of axioms can also be *independent*; this relationship exists if no
   axiom in the collection can be proved by using the other axioms of the collection. A single axiom which cannot be proved from the others is also said to be

---

* For example, the integers are ordered by "less than" in such a way that (1) if
two integers are unequal, one is less than the other, and (2) if one integer is less
than a second and the second is less than a third, then the first is less than the
third. Such an ordering is called a *total ordering* and can be studied quite apart
from the integers.

independent. The usual proof for independence of an axiom is an actual example of an object which satisfies all the axioms except that one. Prove that Axioms 1 through 4 are independent. Explain why an example which satisfies all the axioms except one assures us that the independent axiom cannot be proved from the others.

7. Decide which of the following sets of axioms are consistent. Determine which axioms in each set are independent.

    a) *Axiom 1. Any group contains exactly seven spiders.*
       Axiom 2. Any spider contains exactly four groups.

    b) *Axiom 1. Any two points lie on exactly one line.*
       Axiom 2. Any two lines lie on exactly one point.
       Axiom 3. There are exactly three points.

    c) Axioms 1 and 2 of (b), together with a new axiom.
       Axiom 3. There are exactly four points.

    d) Axioms 1 and 2 of (b), together with a new axiom.
       Axiom 3. Any three lines lie on at least three points.

1.4 MATHEMATICAL STATEMENTS

We have seen that underlying every mathematical system are its axioms, or fundamental assumptions. The mathematical system itself is nothing more nor less than the fundamental assumptions together with the logical consequences of these assumptions and any definitions that may arise along the way. That is, in any mathematical system we have, besides the terms which are left undefined and the axioms which are left unproved, terms which are defined (ultimately, however, by means of the undefined terms) and the theorems which we can prove (ultimately from the axioms). Any mathematical statement we make must be made in the context of one or more mathematical systems. This again is because no purely mathematical statement can be viewed in relation to the real world, but can only be judged in relation to some system constructed by mathematicians.

In other words, a mathematical statement can only be proved or disproved in relation to some set of axioms and definitions. A statement may be true in one axiomatic system and false in another. For example, in Euclidean geometry the sum of the interior angles of a triangle is equal to two right angles. This proposition, however, is quite false in spherical or other non-Euclidean geometries.

Since mathematical statements are like verbal statements, they can take a great variety of grammatical forms. The distinguishing feature of a mathematical statement is that it must be considered entirely within the context of an axiomatic system. From this fact we can draw several important conclusions.

First, we cannot appeal to anything outside that axiomatic system to help us determine the truth or falsity of any statement made within the context of the system.* We may, for example, be able to prove from a certain set of axioms that all men are elephants. That is, in a particular axiomatic system the statement "All men are elephants" either follows in a logically valid manner from the axioms, or is itself one of the axioms. We may know empirically that such a statement is ridiculous and totally false in the "real" world. But in dealing with a mathematical system, we must remember that the axioms determine the "real" world of the system, hence our appeal to experiential data is mathematically meaningless. In mathematics, intuition and experience may be a good guide to what we might expect to hold true, but unless we can supply for a statement a proof that is based on the axioms, i.e., the basic assumptions, of the system in which our statement is intended to be true, we cannot definitively speak of any statement as being true.

Second, since any mathematical statement is made within the context of an axiomatic system, the statement implicitly assumes both the axioms of the system and whatever we have already shown to follow logically from those axioms. Thus when we attempt to prove a mathematical statement, we may use all the machinery already known to exist in the system in which we want to prove the statement. The statement of Theorem 1.1 is made with the implicit supposition that Axioms 1 through 4 hold. Note that the theorem also gives an added condition which may be used in proving the conclusion, namely: A flock contains exactly three blinks; i.e., there are exactly three blinks. We can use this condition, together with the axioms (and definitions too, of course) to prove the conclusion: Each sark is common to exactly two blinks. Thus in Theorem 1.1, there is an assumption above and beyond what is contained in the axioms and what has been previously proved. Such an added assumption is called a *hypothesis*. That is, if some conclusion does not necessarily follow from the axioms of the system alone, but *would* follow if certain other conditions were also satisfied, the extra conditions are hypotheses, and fulfillment of the hypotheses causes the conclusion to be true. For example, in the context of Euclidean plane geometry, it is not always true that any triangle is equilateral. That is, the statement "Any triangle is equilateral" is false, but the statement "If a triangle is equiangular, it is equilateral" is true. In discussing any statement then we must consider both what we already know about the system in which the statement is made and all hypotheses which are attached to the statement.

Third, since any mathematical statement must be discussed in the context of an axiomatic system, it must be stated in terms of that system. To illustrate this point, suppose you are having a spirited discussion with

---

* Of course we must use the laws of logic which are always presupposed.

someone about the merits of some particular politician and suddenly your friend interjects, "I saw a very funny Donald Duck cartoon last week." You would then probably look at your friend a bit oddly and ask, "What has that to do with our discussion of the politician?" If such total irrelevancies in ordinary conversation are bad, they are even worse in mathematics. Although you might not see the point in your friend's statement, at least you understood what it meant. It did have meaning. In an axiomatic system, statements have meaning only if they are expressed in the terms of the system, i.e., by using the undefined terms of the system, terms defined using the undefined terms, or certain terms which are somewhat "metamathematical," that is, more or less basic to our thought processes, such as "equal," "is," etc. Thus, if we were discussing Euclidean plane geometry, then the statement "I saw a very funny Donald Duck cartoon" would not only be irrelevant; it would be completely meaningless. It would be as if someone threw a quotation in Russian at someone who understood only English.

In order to prove a mathematical statement made within the framework of some axiomatic system, we may use the axioms and all that we have validly proved from the axioms. We may also use any additional hypotheses peculiar to the given statement. The proof of the statement is a demonstration according to the laws of logic that the statement is a necessary consequence of what we already know holds true in the system. Thus a mathematical system is essentially pyramidal with the axioms of the system at the base of the pyramid. This pyramidal structure distinguishes mathematics from many other academic disciplines, and makes mathematics rather difficult for many students. In history, for example, one may know a great deal about the reign of Henry VIII without knowing much about Henry VII. If history had the structure of a mathematical system, this would be impossible. Of course, a mathematician might know more about one branch of mathematics than another, but the various branches have distinct axiomatic systems; for the most part, they are different pyramids. Within one branch, a mathematician can study the axiomatic system and develop it in many distinct ways, but always what he can know depends on what he already knows, and everything, no matter how complex, rests ultimately on the axioms.

Much has been made of proving something "validly," in accordance with the laws of logic. What are these laws of logic and on what foundation do they rest? The laws of logic are the laws of valid reasoning, or inference. Logic is an axiomatic system just like any regular mathematical system, except for the very important difference that logic is supposedly the science of correct reasoning and hence underlies, and is presupposed in, all mathematical systems. The reader may conclude, and rightly, that logic is as much philosophy as mathematics; in any university, one will find logicians in both the departments of mathematics and philosophy. As in anything that is philosophical, there is controversy, and there is some disagreement among logicians as to what really constitutes the laws of logic.

If there is debate as to what constitutes these laws, then there is some question as to what constitutes valid reasoning, in particular, which mathematical proofs are genuinely valid. The controversy in logic centers on certain statements held by some to be proved and by others to be undecided. This dispute is quite profound and we cannot discuss it here, but we have introduced the matter to point out that even at the foundation of reason there are unsettled issues. It is even more interesting to ponder that somewhere in this vast universe there may be intelligent life which "thinks" according to laws that we cannot even begin to imagine, and perhaps, by means of its reasoning process, this life has made far greater strides than we.

EXERCISES 1.4

1. Discuss possible procedural differences that might be found between the mathematician and each of the following:

    a) historian   b) biologist   c) philosopher
    d) politician   e) lawyer

2. In each of the following six examples, the letters A and B represent statements. Which of the examples are always true, regardless of what A and B actually mean?

    a) If A is true, then either A or B is true.
    b) If A is false, then A and B are not both true.
    c) If A is not true, then A is false.
    d) If A is true, then B is also true. But B is not true, hence A cannot be true.
    e) If A is true, then B is true. But A is not true, hence B is also not true.
    f) If A is not true, then the statement "A or B is true" is false.

3. Each of the following statements is to be taken in the context of Euclidean plane geometry. For each statement, indicate the conclusion and any hypotheses. [*Note:* A statement which includes some hypothesis need not be given in an "If . . . , then . . ." form; e.g., "Any equilateral triangle is isosceles" is equivalent to "If a triangle is equilateral, then it is isosceles," hence we have the hypothesis, "The triangle is equilateral."]

    a) If opposite sides of a quadrilateral are parallel, then the quadrilateral is a parallelogram.
    b) Two triangles are similar whenever their corresponding angles are congruent.
    c) If a quadrilateral is a parallelogram, then opposite sides are congruent.
    d) Any two congruent triangles have the same area.
    e) All squares are parallelograms.

1.5 NOTATION

Much notation will be introduced during the course of this book, so we should pause here to discuss some general aspects of notation. Notation is an aid to expression because it compresses lengthy verbal statements into concise symbols. Excessive use of notation tends to make material harder for the student to master, but some symbols are essential to an understanding of mathematics.

Notation, being a tool, a shorthand, will be introduced in this text whenever appropriate. Indeed, we have already used letters to denote the statements in Exercise 2 of Section 1.4. Our policy will be to use notation when it aids in understanding the ideas it is intended to convey. As a rule, this text will be relatively sparing in its use of notation.

The use of symbols to denote objects. The reader has in his previous mathematical experience encountered such problems as:

1. Solve the equation $x + 2 = 5$;

or

2. Solve the equation $x^2 + 5x + 6 = 0$.

In Exercise 2 of Section 1.4, the letters A and B were used to denote statements. It is often convenient to allow a symbol, generally a letter of the alphabet, to represent some object, or member of a collection of objects, when there is no other uncomplicated way of specifying the object.

In (1) we are really saying: "Let x denote any number which when added to 2 gives 5; find out what values x can have." Similarly, (2) means: "Let x denote any number such that the square of that number added to five times that number added to 6 is equal to 0; find the values that x can have." In the first instance, x can have only the value 3, while in the second instance, x can be either -2 or -3. We could, of course, get around using x (or another symbol) in (1) by writing something like: "There is a number which when added to 3 gives 5; find that number." But it should also be clear that as the mathematical relationships become more involved, so would our verbal expressions of them. For example, to avoid the use of x in (2), we would have to say: "There are numbers such that given any one of these numbers, the square of the number added to five times the number plus 6 more is equal to 0; find all such numbers." This is certainly more cumbersome than the expression of (2) in terms of x.

Or we might say, "Let y be a student at Harvard." In such an instance, y would refer to an arbitrary student at Harvard, i.e., any student, rather than one in particular. If we wished to achieve the same effect without using a symbol, we would have to keep repeating some phrase like "an arbitrary student at Harvard," or "any student at Harvard." This would not be impossible, but it would be inconvenient.

1. *Do not use a letter or other symbol to refer to some specific object which you can characterize more clearly in another manner.* For example, do not say, "Let x be a student at Harvard" when you mean to refer to one particular student to the exclusion of all others. In such an instance, it would be more appropriate to refer to the student by name or other characterizing feature, e.g., the captain of the football team. We generally use symbols to denote objects when the objects are "unknowns" or arbitrary members of some collection of objects.

2. *As a rule, if it is permissible to represent an object by a letter or symbol, then any letter or symbol is as good as any other symbol for doing the job.* However, this statement must be tempered by the following stipulation.

3. *Some symbols through general usage have come to designate some particular thing*, e.g., 2 denotes the integer two, and π denotes the ratio of the circumference of a circle to its radius. Symbols with a commonly recognized meaning should not be used to designate any object other than that which they customarily designate, except in a context where no confusion could possibly result. For example, we might use π to denote an object in some discussion in which no circles are involved. Finally, it would obviously be unwise to make such statements as "Let x be any letter," or "Let 5 be any number."

4. *Once a symbol has been introduced to represent an object, that symbol should be used consistently to represent only that object.* For example, if Q is allowed to represent any statement, it would be unwise during the course of the same discussion to allow Q to stand for any building as well. Of course, if we wish to designate several of the objects that our symbol represents, we would select other symbols to stand for the similar objects. In Exercise 2 of Section 1.4, both A and B are allowed to stand for statements so that we can discuss two statements simultaneously.

The symbol $=$(df). One notation will be used repeatedly in this text. The symbol $=$(df) will be used in place of $=$, the usual equality sign, to indicate the direct use of a definition where this notation is useful for clarity. That is, $A =$(df)$ B$ means that A and B are the same by definition. All mathematics, of course, ultimately flows from the manner in which we define things, but $=$(df) will indicate that a definition is merely being repeated, or is being stated for the first time.

For example, the sentence "John is the leader of the band" could be rewritten "John $=$ leader of the band." But since being the leader of the band is more or less incidental to John's being John (that is, John is not the leader of the band in such a manner that if he ceased to be leader of the band, he would cease to be John), it would be improper to write "John $=$ (df) the leader of the band." On the other hand, "A thesaurus is a book of synonyms" is the definition of a thesaurus, hence "Thesaurus $=$ (df) book of synonyms" would be quite correct.

EXERCISES 1.5

1. Decide whether $=$ (df) may properly be used to express "is" in each of the following statements:

   a) A triangle is a three-sided polygon.

   b) An equilateral triangle is a triangle with three equal angles.

   c) A pentagon is a polygon with five sides, the sum of whose interior angles is greater than a straight angle.

   e) The area of a triangle is half the product of its height and its base.

   f) Logic is the science of valid inference.

   g) Two plus one is three.

   h) The weather is warm.

2. Write out each of the following without using x, y, or z to denote objects.

   a) Let x and y be politicians. Then x is probably ambitious and regards y with distrust.

   b) Let x and y be any two numbers such that $y = 2x + 5$ and let $z = 3y + 4$. Then $z = 6x + 19$.

   c) There are no positive integers x, y, and z such that $x^3 + y^3 = z^3$.

   d) If x is true, then y is true; but if y is false, then x is false.

3. Write out each of the following statements using symbols to denote objects. Preferably, do this in such a way that the statement becomes more concise.

   a) If one triangle is congruent to a second triangle and the second triangle is congruent to a third triangle, then the first triangle is congruent to the third triangle.

   b) Let one number be found from another number by multiplying that number by 3 and adding 7.

   c) There is a number which when multiplied by 8 and added to 7 gives 4; find that number.

1.6 ON RIGOR

Mathematical rigor is the requirement to be logically precise. Like most good things, rigor can be carried to extremes. Overpreciseness tends to dampen the spirit, dull the imagination, and produce boredom and frustration rather than an admiration for the rational beauty of mathematics. On the other hand, too little rigor leads to carelessness and error as well as to a feeling that mathematics is merely a form of hocus-pocus.

In any book a certain amount of literary license is permitted the author. In a textbook, however, no license can be permitted which might confuse or misinform the reader. At times, the author has felt it permissible, and even advisable, to imply certain statements rather than to state them explicitly when these statements seemed obvious enough. Also certain words or phrases which are merely slight variations of words or phrases previously defined are used without explicitly defining the variations when the meaning is obvious from the context in which they are used. At times certain details are omitted from a proof because these details should be obvious to any reader who has even vaguely understood what has gone before. To define every term with all its variations and supply every single detail would make for a prohibitively long and dry book.

Nevertheless, if the reader finds that any statement in the book seems mysterious, he should immediately try to clarify it before proceeding further. Sometimes it is also valuable to scan an entire section before getting too involved with single words and phrases. This practice will give the reader some indication of where the section is leading and where there are helpful examples.

Let the reader be of good heart and open mind, neither too picky nor too gullible. Having made these points, we now advance to the study of logic.

chapter two

Introduction to Logic

2.1 NECESSARY AND SUFFICIENT CONDITIONS

Sometimes whenever one condition occurs, another condition always occurs with it. For example, we might assert, "If the sun is shining, the air is warm." Thus, whenever we know that the sun is shining, we also know that the air is warm. The statement

If the sun is shining, the air is warm

is equivalent to (says exactly the same thing as) each of the following:

Whenever the sun shines, the air is warm.
All the time the sun shines, the air is warm.
"The sun is shining" implies that the air is warm.
The sun is shining only if the air is warm.

We are thus justified in saying that if we can prove that the sun is shining, we can prove that the air is warm. Again, in order to show that the air is warm, it suffices to show that the sun is shining. We say that in this situation, the sun's shining is *sufficient* for the air to be warm. Note, however, that the air might still be warm even if the sun is not shining; that is, we have not asserted that if the air is warm, then the sun must be shining.

Example 1. The statement "All men are human" is true. But the statement "All men are human" is equivalent to saying "If something is a man, it is human"; thus, being a man is a sufficient condition for being human.

Suppose now we assert, "The air is warm only if the sun is shining." This statement is equivalent to each of the following:

Whenever the air is warm, the sun is shining.
If the air is warm, the sun is shining.
"The air is warm" implies that the sun is shining.

In this case we are saying that warmth cannot occur unless the sun is shining; that is, in order for the air to be warm, it is *necessary* that the sun be shining. Note that here not only is the sun's shining necessary for the air to be warm, but the air's being warm is a sufficient condition for the sun to be shining. For if the air can be warm only when the sun is shining, and we know that the air is warm, then we also know that the sun is shining.

Example 2. The statement "All men are smart" is equivalent to the statement "Something is a man only if it is smart." Thus, being smart is a necessary condition for being a man. It is not sufficient, however, since we have not precluded the possibility of there being smart things other than men.

More formally, we make the following definition.

Definition 2.1. *Condition A is a sufficient condition for condition B if B always occurs when A occurs. That is, we have "If A, then B." Condition A is a necessary condition for B if B cannot occur unless A also occurs. That is, "B only if A." If A occurs when and only when B occurs, then A is said to be a necessary and sufficient condition for B (that is, "A if and only if B").*

Note that if A is a necessary condition for B (that is, B only if A), then B is a sufficient condition for A (that is, if B, then A) for if B can occur only when A occurs, then if B occurs, A must occur as well. On the other hand, if A is a sufficient condition for B, then B is a necessary condition for A.

Example 3. The statement

1. It rains only when it pours

is the same as "If it rains, it pours." On the other hand, the statement

2. If it pours, it rains

is the same as "It pours only when it rains." In (1), pouring is a necessary condition for raining, and raining is a sufficient condition for pouring. In (2), pouring is a sufficient condition for raining, and raining is a necessary condition for pouring. If we say

3. It rains when and only when it pours,

that is, if we combine (1) and (2), then raining would be a necessary and sufficient condition for pouring. Likewise, pouring would be a necessary and sufficient condition for raining.

The reader should be careful to note that if A is a sufficient condition for B, then even though B always occurs whenever A occurs, it is possible that B might occur without A. Thus, if snow is a sufficient condition for the air to be cold, i.e., if it is snowing, the air is cold, it is still possible to have cold air without snow. Similarly, if A is a necessary condition for B, it is possible for A to occur without B. Thus, if snow is a necessary condition for cold air, i.e., the air is cold only when it snows, it is possible to have snow even when the air is not cold.

Another important observation is that if A is a necessary and sufficient condition for B, then B is also a necessary and sufficient condition for A. For if A is a necessary condition for B, then B is a sufficient condition for A; and if A is a sufficient condition for B, then B is a necessary condition for A. Therefore if A is necessary and sufficient for B, B is also necessary and sufficient for A. If A is a necessary and sufficient condition for B, that is, A if and only if B, then A and B always occur simultaneously; it is impossible for one to occur without the other.

Since any "only if" statement can be converted into an "If . . . , then . . ." statement (e.g., "I will wear my coat only if it is cold" becomes "If I wear my coat, then it is cold"), we need to consider only how to handle "If . . . , then . . ." statements as far as mathematical proofs and logic are concerned. In proving any statement of the form "If A, then B," A is a hypothesis and B is the conclusion. For example, in proving "If two sides of a triangle are congruent, then the angles opposite these sides are also congruent," the hypothesis is "Two sides of a triangle are congruent" and the conclusion is "The angles opposite the congruent sides are congruent."

Suppose we wish to prove the statement, "A triangle is equilateral if and only if it is equiangular." This statement is equivalent to two "If . . . , then . . ." statements: "If a triangle is equilateral, it is equiangular" and "If a triangle is equiangular, it is equilateral." Thus, the proof will consist of two parts, one part to prove each "If . . . , then . . ." statement. In general, the proof of a statement of the form "A if and only if B" consists of one part with A as the hypothesis and B as the conclusion and a second part with B as the hypothesis and A as the conclusion. Looking at it another way, "A if and only if B" means that A is a necessary and sufficient condition for B; hence any proof of such a statement would have to show two things: (1) A is a necessary condition for B, and (2) A is a sufficient condition for B.

Example 4. In order to prove "A quadrilateral is a parallelogram if and only if opposite sides are congruent," we must prove

4. If a quadrilateral is a parallelogram, then opposite sides are congruent

and

5. If the opposite sides of a quadrilateral are congruent, then it is a parallelogram.

Statement (4) tells us that being a parallelogram is a sufficient condition for the opposite sides of a quadrilateral to be congruent, while (5) tells us that being a parallelogram is a necessary condition for the opposite sides of a quadrilateral to be congruent. A complete proof of the original "If and only if" statement really requires two proofs, a proof of (4) and a proof of (5).

EXERCISES 2.1

1. In each of the following statements, determine whether one condition is necessary for another condition, sufficient for another condition, or both necessary and sufficient.
   a) If I do this, I am foolish.
   b) I will come if it does not rain.
   c) I will come only if it does not rain.
   d) When I see you, I will shout.
   e) The air is warm only when the sun shines.
   f) Whenever the phone rings, I run to answer it.
   g) I will wear a coat if and only if it snows.
   h) When I feel happy, I laugh; and when I laugh, I am happy.
   i) All men can read. (One condition here is being able to read, and the other is being a man.)
   j) Man is a rational animal. (The "is" should be interpreted as meaning "is by definition.")

2. Convert all of the following into "If ... , then ..." statements.
   a) I will go only if you say so.
   b) When I think about study, I want to play.
   c) You will find it when you look for it.
   d) No man is an elephant.
   c) In order for two triangles to be similar, it suffices for the angles of one to be congruent to the angles of the other.
   g) It is necessary for you to eat in order to live.
   h) All ambitious people work hard.

2.2 NEGATIONS

If anyone were asked what is the *negation** of the statement

6. It is raining now,

he would probably answer correctly

7. It is not raining now.

We note that if (6) is true, then (7) is false. Similarly, if (6) is false, then (7) is true.

However, the simple insertion of a "not" in some statement is generally not sufficient to obtain a negation of that statement. For example, consider the statement "Some men are smart." We might suspect that the negation of this statement is "Some men are not smart." Observe, however, that both statements, the original and the supposed negation, can be true simultaneously; that is, it may well be that some men are smart and some men are not smart.

Definition 2.2. If A is allowed to represent any statement, then a negation of A is any statement B which is false whenever A is true, and which is true whenever A is false.

In other words, if B is a negation of A, then:

8. If A is true, then B is false,

and

9. If A is false, then B is true.

Thus, B is a negation of A if the falsity of A is a sufficient condition for the truth of B, *and* the truth of A is a sufficient condition for the falsity of B.

The correct negation then of "Some men are smart" is "All men are non-smart." For if some men are smart, then it is false that all men are non-smart; and if it is false that some (i.e., at least one) men are smart, then it is true that all men are non-smart.

The fact that any statement and its negation cannot simultaneously be true means that if we show that some statement is true, then any negation of that statement must be false. That is, a proof of any statement amounts also to a disproof of any negation of that statement. Thus, if we show that some man is non-smart, e.g., by producing an actual example of someone who is non-smart, we have proved the statement

---

* Sometimes called the *contradictory* or *denial*.

"Some men are non-smart" and disproved the statement "All men are smart" as well.

We remark that not only is (7) the negation of (6), but (6) is the negation of (7) too. In general, we may ask: If B is a negation of A, is A also a negation of B? We first observe that if B is a negation of A, then (8) and (9) hold. If A is also to be a negation of B, then we will have to have:

10. If B is true, then A is false,

and

11. If B is false, then A is true.

Let us suppose then that B is a negation of A and that B is true. If it were possible for A to be true when B is true, then statement (8) could not hold; for (8) says that when A is true, B is always false. Thus, if B is true, A is false; hence (10) holds. If B and A could both be false simultaneously, then (9) would be contradicted; therefore if B is false, A must be true. Consequently, (11) also holds true. We have thus shown that if A is a negation of B, then B is also a negation of A.*

Example 5. A negation of

12. The earth is round

is

13. The earth is not round.

For if (12) is true, then (13) is false, and if (12) is false, then (13) is true. Likewise, (12) is a negation of (13). The statement

14. The earth is a cube

is not a negation of (12). If (12) is true, then (14) is false; but if (12) is false, then (14) may or may not be true, i.e., the earth may have some shape other than that of a cube.

We *cannot* say that a statement A is a negation of B if whenever A is true, B is false, and whenever B is true, A is false; for this is precisely the situation we have with (12) and (14).

Example 6. The negation of "It is possible that it will rain today" is not "It is possible that it will not rain today" since both statements are quite

---

* The technique used in this demonstration was the so-called *proof by contradiction*, which will be studied at length in Section 3.5.

compatible, that is, both can be true simultaneously. A genuine negation would be "It is not possible that it will rain today."

To sum up then, a statement B is a negation of a statement A if whenever A is true, B is false, and whenever A is false, B is true. If B is a negation of A, then A is a negation of B. If B is a negation of A, then a proof of A is at the same time a disproof of B, i.e., a demonstration that A is true is also a demonstration that B is false; likewise, a proof of B is a disproof of A.

EXERCISES 2.2

1. In each of the following exercises, two statements are given. Determine whether the second is a negation of the first.
   a) I am thirsty.
      I am not thirsty.
   b) All elephants are red.
      One elephant is not red.
   c) Some horses are black.
      Some horses are brown.
   d) All horses are black.
      Some horses are brown.
   e) If the sun is shining, it is warm.
      If the sun is not shining, it is warm.
   f) If I am right, you are wrong.
      If I am wrong, you are right.
   g) No man is an elephant.
      Some men are elephants.
   h) All fire engines are red.
      All fire engines are yellow.
   i) Sometimes I am wrong.
      All of the time I am right.

2. Determine a negation for each of the following statements.
   a) It is mild out today.
   b) All my cows are in the pasture.
   c) Some of my cows are in the pasture.
   d) Only one of my cows is in the pasture.
   e) No two men are alike.
   f) If no one else cares, then I don't either.
   g) Whenever it rains, I wear my coat.

3. Suppose some statement B is a negation of a statement A. We have already seen that A is a negation of B. Suppose that statement C is a negation of B. That is, we have B is a negation of A and C is a negation of B. Prove that A is true if and only if C is true.

2.3 MATHEMATICAL DISPROOFS

Many times we are confronted with statements which we really do not know whether to believe or not. For example, at the beginning of a criminal trial the prosecutor may state without reservation that the defendant is guilty, while the defense attorney will insist on the innocence of his client. A fair juror will wait to weigh all the evidence before coming to a decision.

Often, too, a statement may have some truth in it, even when it is false. The old saw "Money is the root of all evil" is based on the well-known phenomenon that greed makes people do strange things; but, of course, money can also be a genuine good if gained and used properly. The proverb is false, but has a grain of truth.

In mathematics a statement is either provable in the context of the axiomatic system in which it is made, or it is not provable. Of course, if it is provable, it is true (since this is what we mean by truth so far as mathematics is concerned). But suppose we cannot find a proof for the statement. What then?

It is entirely possible that the statement is true even if we cannot prove it. It may be merely a lack of knowledge which holds us back. It is possible that when the problem is better understood or some new approach is available, a proof will be forthcoming.

Example 7. The great French mathematician Pierre de Fermat (1601–1665) believed that he had found a proof for the following theorem: There are no positive integers x, y, and z such that $x^n + y^n = z^n$ for any integer n larger than 2. For example, there are no positive integers x, y, and z such that $x^5 + y^5 = z^5$. Unfortunately, Fermat never wrote his supposed proof down, and it is strongly suspected nowadays that he did not have a valid proof at all. This theorem, known as *Fermat's Last Theorem* (since Fermat died shortly after proposing it), has been tackled by virtually every would-be mathematician since the time of Fermat but no proof has as yet been found. The theorem is known to be true, however, for n as large as 10,000 and no one has as yet found an integer n larger than 2 for which there are positive integers x, y, and z such that $x^n + y^n = z^n$. In sum, all evidence suggests that Fermat's Last Theorem is true, but no one can be absolutely certain until a proof is found.

We see then that the absence of a proof does not necessarily mean that a statement is false. But suppose we do wish to disprove a statement, that is, to show that a statement is really false. As we saw in this last section, one thing we might do is to prove a negation of the statement. For example, to prove "All triangles are isosceles" is false, it suffices to prove that "Some triangle is not isosceles" is true.

Example 8. Suppose we wish to disprove that $x^2 + y^2$ is always an even integer whenever x and y are integers. That is, we wish to show that the statement "The sum of the squares of any two integers is an even integer" is false. We can do this by proving the negation of this statement, i.e., "There are two integers x and y such that $x^2 + y^2$ is an odd integer." If we let $x = 2$ and $y = 3$, then we obtain $x^2 + y^2 = 4 + 9 = 13$, which is not even. We have thus disproved the statement we wished to disprove by showing that its negation is true.

Example 9

15. Every positive integer can be expressed as the sum of the squares of three or fewer integers; that is, if n is any positive integer, then there are integers x, y, and z (one or more of which may be zero) such that $x^2 + y^2 + z^2 = n$.

Now

$$1 = 1^2(+0^2 + 0^2), 2 = 1^2 + 1^2, 3 = 1^2 + 1^2 + 1^2,$$
$$4 = 2^2, 5 = 2^2 + 1^2, 6 = 2^2 + 1^2 + 1^2.$$

The number 7 cannot be expressed as the sum of three squares, but 7 can be expressed as the sum of four squares, namely,

$$7 = 2^2 + 1^2 + 1^2 + 1^2.$$

Statement (15) is false. But at the same time, believing that (15) is close to the truth, we might ask whether each positive integer is the sum of at most four squares. The latter is in fact the case.

Actually, when a conjecture is well founded, even when it is false, it may be extremely difficult to disprove. Note that although the disproof of any statement is really a proof of any of its negations, any negation may be a much weaker statement than the original statement. For example, a negation of "All men are green," a statement referring to all men without exception, is "There is at least one man who is not green." To prove this negation, we need only supply one example of a non-green man; but proof of this negation is not nearly as strong as proving "No man is green," a statement which, like the first, refers to all men without exception.

An example used to show that some statement is false is called a *counterexample*. We negate a statement by finding an instance where the statement claims to be true but is not. The reader should not confuse the procedure of disproving some statement by producing a counterexample, with disproving a statement by *contradiction*, a technique to be discussed in a later section.

Remember the following principles. They are often violated.

1. *In order to be true, a statement must hold in every instance in which it claims to hold.* Thus a statement like "All triangles are isosceles" states that given any triangle whatsoever, that triangle is isosceles. We can therefore disprove this statement by finding just one example of a triangle that is not isosceles. (We would obviously, however, not be able to prove the statement by finding one example of a triangle which is isosceles.)

Note that the statement "If it is snowing, I won't go" states that if the condition of snowing is satisfied, then I won't go. A conclusion need not be true if its hypotheses are not satisfied. For example, if the statement "If it is snowing, it is cold" is true, it does not necessarily have to be either cold or snowing, but if it is snowing, then it must be cold. And if it is not snowing, it needn't be cold. If we can find but one instance when it is snowing and is not cold, we will have proved the statement false.

2. *Even a false statement may be true in special instances, or at least be partly true*, e.g., see Example 9. Demonstrating that a statement is true for one or two special instances then does not constitute a proof of the statement unless the statement only claims to be true for one or two special instances. For example, "There is an isosceles triangle" would be proved by actually producing an isosceles triangle. A statement must always be proved in such a way that all situations which the statement claims to cover are taken into account. If we wish to prove "The sum of the interior angles of any triangle is equal to two right angles," then we must seek a proof that accounts for all triangles and not just one or two, or even a special type of triangle. However, in order to prove the statement false, it would be necessary to find only one triangle whose interior angles do not add up to two right angles. We should note, however, that sometimes we can gain deeper insight about a statement if we consider a particular situation to which the statement applies, even though showing that the statement is true in that particular instance does not itself constitute a proof of the statement. Such an application may direct us to a valid proof after we have considered it thoroughly.

EXERCISES 2.3

1. In order to disprove each of the following statements, what statements would we actually prove?
   a) All men are smart.
   b) Socrates is a man.
   c) If the sun is shining, it is warm.
   d) The sum of any two even integers may be odd.
   e) Some triangles are isosceles.
   f) Nobody knows the trouble I've seen.

2. Find a counterexample for each of the following statements and thus disprove them.
   a) If x is any number, then $x < x^2$.

b) All triangles are isosceles.

c) The sum of the squares of any four integers, each of which is not zero, is always an even integer.

d) All birds can fly.

e) If the four sides of one quadrilateral are congruent to the four sides of another quadrilateral, then the two quadrilaterals are congruent.

3. Suppose we wish to prove a statement A. Is finding a counterexample for a negation of A equivalent to proving A? Explain.

4. Can a statement such as "Some men are silly" be disproved by finding an example of a man who is not silly? Discuss what type of statement we would have to have in order to be able to disprove it by finding a counterexample.

2.4 CONJUNCTION AND DISJUNCTION

Two of the most common words in the English language are "and" and "or." Although grammatically we classify both of these words as conjunctions, their meanings are, of course, quite different. The reader has been using these two conjunctions virtually all his life and is no doubt aware of the way they are used in everyday conversation, which indeed is also the way they are used in mathematics. But we will now investigate the properties of these simple words both to clarify our thinking and because of their importance in many proofs.

To say

16. John is a good student and a gentleman

is equivalent to saying both

17. John is a good student

and

18. John is a gentleman.

Similarly, if we say, "I like meat and cheese," we have really affirmed two simple statements: (1) "I like meat," and (2) "I like cheese."

If we declare that (16) is true, then both (17) and (18) must also be true. Likewise, if both (17) and (18) are true, then (16) is true as well. On the other hand, if either (17) or (18) is false, even if the other happens to be true, then (16) is false. Likewise, if (16) is false, then either (17) or (18), possibly both, is false.

More formally, if we have some statement of the form "A and B," then if "A and B" is true, both A and B are true separately. Likewise, if both A and B are true separately, then "A and B" is true. If either A or

B is false, then "A and B" is also false (since it could only be true when A and B are true simultaneously). Thus, we can make each of the following statements about a statement of the form "A and B":

19. a) If A and B are each true, then "A and B" is true.

    b) If "A and B" is true, then A and B are each true.

    c) If either A or B is false, then "A and B" is false.

    d) If "A and B" is false, then either A or B is false.

Example 10. If the statement "It is raining and (it is) warm" is true, then both "It is raining" and "It is warm" are true. On the other hand, if it is not raining, or it is not warm, then "It is raining and warm" is false.

Combining two statements by "and" is called *conjunction*. At times conjunction may also be effected by grammatical devices equivalent to "and," e.g., "John is a good student, but he is also a gentleman" is equivalent to (16).

The joining of two statements by "or" is called *disjunction*, e.g., "John is a good student or a good athlete." At times, disjunction may also be effected by grammatical devices equivalent to "or"; e.g., "He lived at one of the two addresses you mentioned."

If we assert

20. John is a good student, or a good athlete,

we would intend one of the following:

21. John is a good student, or a good athlete, but not both,

or

22. John is a good student, or a good athlete, possibly both.

Statement (21) is an example of *exclusive disjunction*, while (22) is an example of *inclusive disjunction*. If (21) is true, then one and only one of the following is true:

23. John is a good student.

24. John is a good athlete.

If (22) is true, one of (23) and (24) is true, and possibly both (23) and (24) are true. Regardless of which interpretation of (20) we take, if (20) is true, then at least one of (23) and (24) must be true; moreover, if (20) is false, then both (23) and (24) must be false.

We shall always consider disjunction to be *inclusive* unless a particular disjunctive statement is specified as exclusive.

In dealing with any statement of the form "A or B," we may say the following:

25. a) If A and B are each true, then "A or B" is true if we assume that the disjunction is inclusive.

    b) If "A or B" is true, then at least one of the statements A and B is true. (This holds regardless of which way we interpret "or," that is, whether we take it to mean exclusive or inclusive disjunction.)

    c) If either A or B is false, then "A or B" is true so long as A and B are not false simultaneously.

    d) If "A or B" is false, then both A and B are false. Likewise, if A and B are both false, then "A or B" is false.

Example 11. If the statement "It is raining, or snowing" is true, then either it is raining, or it is snowing. But if "It is raining, or snowing" is false, then it is neither raining, nor snowing.

In order to prove any statement of the form "A and B," we would have to prove both A and B. For example, in order to prove "Any equilateral triangle is isosceles and equiangular," we must prove both "Any equilateral triangle is isosceles" and "Any equilateral traingle is equiangular." In order to prove any statement of the form "A or B," we must prove that always *one* of the statements A and B must be true, regardless of whether the "or" is meant exclusively or inclusively. For example, in order to prove "The square of any integer is either zero or positive," we must show that if the square of any integer is not zero, then it must be positive (or, if the square is not positive, it is zero).

EXERCISES 2.4

1. What is wrong with the following argument?
    Each integer is either even or odd. That is, either "Each integer is even" is true, or "Each integer is odd" is true. But it is false that each integer is even, and it is also false that each integer is odd. Thus it must also be false that each integer is either even or odd.

2. Break down each of the following statements into simple statements, as, for example, (16) was broken down into (17) and (18); and indicate whether the simple statements should be combined with "and" or "or."

    a) He is both safe and well.
    b) Sam is either happy or sad.
    c) They are neither happy nor sad.
    d) They are happy, or they are not happy.

e) No one is here, but I don't care.

f) Any angle is acute, right, obtuse, straight, reflex, or round.

g) Neither Harry nor Sue bothered to go to class.

2.5 TRUTH TABLES

Any statement can have one of two *truth values*: true or false.* If A and B are any two statements, we can prepare the following tables from the information given in (19) and (25).

Table 2.1

If A is	and B is	then "A and B" is
true	true	true
true	false	false
false	true	false
false	false	false

Table 2.2

If A is	and B is	then "A or B" is
true	true	true
true	false	true
false	true	true
false	false	false

Tables 2.1 and 2.2 are examples of *truth tables*. A truth table is merely an array which gives the truth value of some statement when the truth values are specified for the statements on which the given statement depends. For example, the truth value of "A and B" depends on both the truth value of A and the truth value of B. Table 2.1 gives the truth value of "A and B" when A and B have various truth values. Similarly, Table 2.2 gives the truth value for "A or B" when the various truth values of A and B are specified.

Another truth table might give the truth value of the negation of A when truth values of A are specified (see Table 2.3).

---

* Not all logicians, however, agree that there are only two truth values, even though virtually all mathematicians assume logic is *two-valued*. In a three-valued logic, for example, a statement might be true, false, or undecidable. Cf. the last paragraph of Section 1.4.

Table 2.3

If A is	then the negation of A is
true	false
false	true

Usually, however, truth tables are presented in a much more concise form than the one used in the tables we have looked at. As a rule, the "If . . . , then . . ." column headings are omitted and only the statement heads the column. For example, in place of "If A is . . ." in Table 2.1, we would write merely A. The truth value "true" is denoted by T, and "false" is denoted by F. In addition to these conventions, the following notation is more or less universally used by logicians:

26. "And," or its grammatical equivalents, i.e., conjunction, is denoted by the symbol $\wedge$, an inverted "v" called a *wedge*. Thus "A and B" could be written "$A \wedge B$." (In some texts conjunction is denoted instead by $\cdot$, i.e., "A and B" would be written $A \cdot B$. We will restrict ourselves to the wedge notation.)

27. "Or" in the sense of inclusive disjunction, or its grammatical equivalents, is denoted by $\vee$, a "v" or inverted wedge. Thus "A or B" would be denoted by "$A \vee B$."

28. The negation of A is denoted by $\sim A$.

We now reconstruct Tables 2.1, 2.2, and 2.3 in the form in which they would usually be given:

Table 2.4 (Table 2.1)

A	B	$A \wedge B$
T	T	T
T	F	F
F	T	F
F	F	F

Table 2.5 (Table 2.2)

A	B	$A \vee B$
T	T	T
T	F	T
F	T	T
F	F	F

Table 2.6 (Table 2.3)

A	$\sim A$
T	F
F	T

Note that Table 2.4 not only shows the truth value of $A \wedge B$, given the truth values of A and B; it also shows the truth value of B when the truth values of A and $A \wedge B$ are given. For example, if the truth value of A is T and the truth value of $A \wedge B$ is F, we see from Table 2.4 that the truth value of B must then be F. Given the truth value of any two of the three statements of Tables 2.4 or 2.5, we can read the truth value of the third statement directly from the table.

The truth tables presented so far have been such that given the truth values of A and B (or just A in the case of Table 2.6), we are able to come up with the truth value of the appropriate statement, $A \wedge B$, $A \vee B$, or $\sim A$, directly, that is, without computing any intermediate truth values. Such direct assessment is not always possible. Let us look at the following examples.

Example 12. Suppose we wish to express the relationships between the truth values of A and B and the statement "$A \wedge \sim B$." (For example, if A is "It is raining" and B is "It is snowing," then "$A \wedge \sim B$" is "It is raining and not snowing.") Given the truth values of A and B, we must first compute the truth value of $\sim B$ before finding the truth value of $A \wedge \sim B$.

29. Suppose A is true and B is true. We see from Table 2.6 that if a statement is true, its negation is false; hence $\sim B$ is false. If we have the conjunction of a true statement (A) with a false statement ($\sim B$), then we see from Table 2.4 that the conjunction is false; hence $A \wedge \sim B$ is false.

Once the truth values of A and B are set, then the truth values of $\sim B$ and $A \wedge \sim B$ are also set. All the relationships between the truth values of these four statements are expressed in the following truth table. Table 2.7 is actually a combination of Tables 2.4 and 2.6. Row 1 of Table 2.7 was explicitly computed in (29).

Table 2.7

A	B	$\sim B$	$A \wedge \sim B$
T	T	F	F
T	F	T	T
F	T	F	F
F	F	T	F

Observe that if the truth values of any three of the statements A, B, $\sim B$, and $A \wedge \sim B$ are given, then the truth value of the fourth statement is not always uniquely determined. Consider rows 1 and 3 of Table 2.7. In these rows, B, $\sim B$, and $A \wedge \sim B$ have the same truth values, but A has value T in row 1 and F in row 3.

Example 13. We can compute a truth table to show that $\sim(\sim A)$ always has the same truth value as A, that is, the negation of a negation of a statement always has the same truth value as the original statement. We explicitly compute the first row of the table in argument (30).

30. If A is true, then $\sim A$ is false. If $\sim A$ is false, then $\sim(\sim A)$ is true (we have used Table 2.6 twice); hence if A is true, $\sim(\sim A)$ is true.

Table 2.8

A	$\sim A$	$\sim(\sim A)$
T	F	T
F	T	F

One very important statement which can be formed from statements A and B is "If A, then B." Thus we want to determine the truth value of "If A, then B" when we are given the truth values of A and B. Looking at A and B as conditions or events, we see that "If A, then B" is equivalent to saying "Whenever A occurs, B occurs as well," which in turn is equivalent to saying "It cannot happen that A occurs but B does not occur," i.e., "It cannot happen that A occurs and $\sim B$ occurs." Thus

31. "If A, then B" is equivalent to "$\sim(A \wedge \sim B)$."

By using Table 2.7, we can construct the following truth table.

Table 2.9

A	B	$A \wedge \sim B$	$\sim(A \wedge \sim B)$
T	T	F	T
T	F	T	F
F	T	F	T
F	F	F	T

The relationship "If ..., then ..." or "implies" is customarily denoted by $\rightarrow$ (although $\supset$ is used in some texts).* Thus "If A, then B" would be written "$A \rightarrow B$." Using (31) and Table 2.9, we have the following truth table for $\rightarrow$.

---

* Certain texts, most notably, *Introduction to Finite Mathematics* by Kemeny, Snell, and Thompson, distinguish between "If ..., then ..." and "implies." "If A, then B" is taken to mean "If the condition A is fulfilled, then B occurs," while "A implies B" means that A logically implies B. We shall give "implies" its broadest possible meaning and consider "If ..., then ..." and "implies" to be essentially equivalent.

Table 2.10

A	B	$A \to B$
T	T	T
T	F	F
F	T	T
F	F	T

Rows 3 and 4 of Table 2.10 express the valid idea that a false statement can be employed to prove any statement at all, either true or false. This idea is common in everyday speech, in such expressions as "If you can sing grand opera, then I'm Napoleon." Row 2 indicates that a true statement cannot imply a false statement, but only true statements (row 1).

EXERCISES 2.5

1. Let A be "The wind is blowing" and B be "The weather is warm." Write in words each of the following statements.
   a) $\sim A$
   b) $A \wedge B$
   c) $A \vee B$
   d) $\sim A \wedge \sim B$
   e) $\sim (A \vee B)$
   f) $\sim (\sim A \wedge \sim B)$
   g) $A \to B$
   h) $\sim A \to \sim B$.

2. Suppose A is true and B is false. Determine the truth value of each of the eight statements in Exercise 1.

3. The first row of Table 2.7 is computed in (29). Compute the other rows of Table 2.7.

4. Produce a truth table for each of the following expressions; the table should show the relationships among the truth values of A and B and the expression. Be sure to include in each truth table any statements intermediate between A and B and the given expression, for example, in Table 2.7, $\sim B$ is a necessary intermediate statement.
   a) $A \vee \sim B$
   b) $\sim A \to B$
   c) $\sim A \to \sim B$
   d) $A \to (A \vee \sim B)$
   e) $\sim (A \to B)$

5. Using the truth tables given in this section, determine the truth value of A, given the truth values in each of the following:
   a) B is true and $A \wedge B$ is true
   b) B is true and $A \wedge B$ is false
   c) B is false and $A \vee B$ is true
   d) B is true and $A \wedge \sim B$ is false
   e) $\sim B$ is true and $A \wedge \sim B$ is false
   f) B is true and $A \to B$ is true
   g) B is false and $A \to B$ is false
   h) $\sim B$ is true and $A \to B$ is false
   i) $\sim B$ is true and $\sim (A \to B)$ is false

6. Let A, B, and C be statements. Construct a truth table showing the relationships among the truth values of A, B, and C and each of the following.

a) $A \wedge (B \wedge C)$ The truth table here might begin

A	B	C	$B \wedge C$	$A \wedge (B \wedge C)$
T	T	T	T	T

b) $A \vee (B \vee C)$ c) $A \rightarrow (B \vee C)$ d) $(A \rightarrow \sim B) \wedge C$

7. For each of the following statements let letters stand for appropriate simple statements and express each statement symbolically. For example, for the statement "If it is snowing, I am skiing or asleep," we would let A be "It is snowing," B be "I am skiing," and C be "I am asleep"; the original statement would then become $A \rightarrow (B \vee C)$.

a) If I am wrong, I will not admit it.
b) If a triangle is not isosceles, it is not equilateral.
c) Every integer is either even or odd.
d) If you ran away, you were either ignorant or foolish.
e) If it is not raining, then I am both seeing and hearing things.
f) Anyone who would do such a thing is not in control of his senses.
g) I watch television only when I am nervous or bored.
h) If I go, I will do the wrong thing, but I want to go anyway.

2.6 LOGICAL EQUIVALENCE. TAUTOLOGIES

Some statements are always true. For example,

32. "Socrates is wise, or Socrates is not wise"

is true regardless of whether Socrates really is wise or not. If we let A be "Socrates is wise," then (32) has the form $A \vee \sim A$.

Now let A be any statement whatever; we can construct truth table 2.11 for $A \vee \sim A$. We thus see that any statement of the form $A \vee \sim A$ is true regardless of the truth value of A. Any statement which is always true regardless of the truth values of the simple statements of which it is composed is called a *tautology*.

Table 2.11

A	$\sim A$	$A \vee \sim A$
T	F	T
F	T	T

We have seen (Table 2.8) that A and $\sim(\sim A)$ have the same truth value for any statement A, regardless of the truth value of A. Thus

33. A has the same truth value as $\sim(\sim A)$

is a tautology. Now $\sim(\sim A)$ may not be precisely the same statement semantically as A; for example, if A is "It is snowing," then "It is not not-snowing" could serve as $\sim(\sim A)$. However, so far as truth value is concerned A and $\sim(\sim A)$ are exactly the same.

Definition 2.3. Two statements are said to be logically equivalent if they always have exactly the same truth values. Put another way, A is logically equivalent to B if A is true if and only if B is true. If A and B are logically equivalent, then we may write $A \equiv B$; that is, $A \equiv B$ denotes the fact that A is logically equivalent to B.

Example 14. If two statements actually say the same thing, then, of course, they are logically equivalent. Thus, from (31) we have

34. $A \rightarrow B \equiv \sim(A \wedge \sim B)$.

In fact, we found the truth values of $A \rightarrow B$ by means of the truth values of $\sim(A \wedge \sim B)$.

Example 15. The following properties of logical equivalence follow directly from Definition 2.3.

35. $A \equiv A$.

36. If $A \equiv B$, then $B \equiv A$.

37. If $A \equiv B$ and $B \equiv C$, then $A \equiv C$.

Property (35) merely says that any statement always has the same truth value as itself. Property (36) states that if A always has the same truth value as B, then B always has the same truth value as A. Property (37) states that if A always has the same truth value as B and B always has the same truth value as C, then A always has the same truth value as C. These three properties should be self-evident to the reader.

If some statement A is true when and only when some statement B is true, then A and B are interchangeable in any proof. Again, $A \equiv B$ essentially means that A is a necessary and sufficient condition for B (Section 2.1). Thus we may substitute A for B or B for A in any statement or argument of which either is a part, without affecting either the truth value of the statement or (as we shall see more clearly in a later section) the validity of the argument.

Example 16. We know that a triangle is equilateral if and only if it is equiangular. Therefore "The triangle is equilateral" and "The triangle is equiangular" are logically equivalent statements. Consequently, in any proof we may use these two statements interchangeably without any loss of validity.

Suppose we are given two statements and wish to know whether they are in fact logically equivalent. We must, of course, establish whether or not the statements always have the same truth values. If the situation lends itself to it, we might set up truth tables for each statement. If the truth values of the two statements match whenever the truth values of the corresponding components of each statement match, then the two statements are logically equivalent. We illustrate this technique in the following example.

Example 17. We will show that $\sim(A \wedge B) \equiv (\sim A \vee \sim B)$. For example, if A is "It is raining" and B is "It is windy," then $\sim(A \wedge B)$ would be "Windy and raining, it is not." (We use this awkward phrasing to avoid the ambiguous "It is not windy and raining.") The expression $\sim A \vee \sim B$ is "Either it is not raining, or it is not windy." It should be clear that in this instance at least the two statements are logically equivalent. Truth tables for $\sim(A \wedge B)$ and $\sim A \vee \sim B$ are given in Tables 2.12 and 2.13. The important thing to note is that when the corresponding truth values for A and B are the same in both tables, $\sim(A \wedge B)$ and $\sim A \vee \sim B$ have the same truth values, hence they are logically equivalent statements.

Table 2.12

A	B	$A \wedge B$	$\sim(A \wedge B)$
T	T	T	F
T	F	F	T
F	T	F	T
F	F	F	T

Table 2.13

A	B	$\sim A$	$\sim B$	$\sim A \vee \sim B$
T	T	F	F	F
T	F	F	T	T
F	T	T	F	T
F	F	T	T	T

EXERCISES 2.6

1. Prove that $\sim(A \wedge \sim A)$ is a tautology. Are any two tautologies logically equivalent?

2. Prove each of the following.
   a) $\sim(A \vee B) \equiv (\sim A \wedge \sim B)$ b) $A \equiv (A \wedge A)$
   c) $A \equiv (A \vee A)$
   d) $A \wedge (B \vee C) \equiv (A \wedge B) \vee (A \wedge C)$
   e) $A \vee (B \vee C) \equiv (A \vee B) \vee C$
   f) $A \vee (B \wedge C) \equiv (A \vee B) \wedge (A \vee C)$

3. Verify Tables 2.12 and 2.13.

4. Let A be "Tom is a good student" and B be "Tom is a gentleman." Write in words each statement in each of the following pairs of logically equivalent statements.
   a) $\sim(A \wedge B); \sim A \vee \sim B$ b) $A \rightarrow B; \sim(A \wedge \sim B)$
   c) $\sim(A \vee B); \sim A \wedge \sim B$

5. Prove that $A \equiv B$ is logically equivalent to each of the following.
   a) $(A \rightarrow B) \wedge (B \rightarrow A)$ b) $(A \wedge B) \vee (\sim A \wedge \sim B)$

6. Prove that $A \rightarrow B$ is not logically equivalent to $A \wedge B$.

chapter three

More about Logic

3.1 CONVERSES AND CONTRAPOSITIVES*

If the statement

1. If it is raining, I am wearing my coat

is true, it is not necessarily true that the statement

2. If I am wearing my coat, it is raining

is true; for I might be wearing my coat, of course, even though it is not raining. Statements (1) and (2) would be logically equivalent, i.e., would always have the same truth value, if and only if "It is raining" is a necessary and sufficient condition for "I am wearing my coat."

More generally, if we have any statement of the form

3. If A, then B, (that is, $A \to B$),

it need not be logically equivalent to

4. If B, then A, (that is, $B \to A$).

In fact, (3) and (4) are logically equivalent if and only if A is logically equivalent to B, that is, if A is a necessary and sufficient condition for B.

Definition 3.1. The statement "If B, then A" is said to be the converse of the statement "If A, then B."

For example, (1) is the converse of (2); similarly (2) is the converse of (1).

From what we have just discussed, we see that if we prove any "If . . . , then . . ." statement, we still cannot be sure about the truth of its converse.

---

* The reader might do well to briefly review Section 2.1 before beginning this section.

Further, if we wish to prove that A is a necessary and sufficient condition for B, then we must prove both "If A, then B" and its converse "If B, then A." For example, if we wish to prove "A triangle is equilateral if and only if it is equiangular," we must prove "If a triangle is equilateral, it is equiangular" and "If a triangle is equiangular, it is equilateral."

If the statement

5. When I sing, I sound awful

is true, then the statement

6. When I do not sound awful, I am not singing

is true. Observe that (5) is a statement of the form $A \rightarrow B$ (with A being "I sing" and B being "I sound awful"); and (6) is a statement of the form $\sim B \rightarrow \sim A$, with the same meanings for A and B as in (5). Actually, not only is (6) true when (5) is true, but (5) is also true whenever (6) is true. That is, (5) is true if and only if (6) is true; hence (5) and (6) are logically equivalent statements.

More generally, we shall show that any statement of the form

7. $A \rightarrow B$

is logically equivalent to

8. $\sim B \rightarrow \sim A$.

Definition 3.2. Statement (8) is said to be the contrapositive of (7). Thus, (6) is the contrapositive of (5).

The contrapositive of "If an integer is larger than 1, it is larger than 0" is "If an integer is not larger than 0, it is not larger than 1."

To show that (7) and (8) are logically equivalent, we produce truth tables for each. The truth table for (7) was given in Table 2.10. The truth table for (8) is as shown in Table 3.1. Since (7) and (8) have exactly the same truth value whenever the corresponding truth values for A and B are the same, (7) and (8) are logically equivalent.

Since (7) and (8) are logically equivalent, they can be used interchangeably. In particular, if we wish to prove a statement of the form of (7), we may, if we find it more feasible, prove a statement of the form (8) instead. For if we show that (8) is true, then since (7) always has the same truth value as (8), (7) will also be true.

Example 1. If we wish to prove "A quadrilateral is a parallelogram if its opposite sides are congruent," we could do so either by proving the statement itself or by proving its contrapositive "If a quadrilateral is not a parallelogram, then its opposite sides are not congruent."

Table 3.1

A	B	$\sim B$	$\sim A$	$\sim B \rightarrow \sim A$
T	T	F	F	T
T	F	T	F	F
F	T	F	T	T
F	F	T	T	T

Example 2

9. A triangle is equilateral if and only if it is equiangular

is equivalent to

10. If a triangle is equilateral, it is equiangular

together with

11. If a triangle is equiangular, it is equilateral.

One way of proving (9) would be to prove (10) and then prove the contrapositive of (11) "If a triangle is not equilateral, it is not equiangular." Of course we could also prove (9) by proving (11) and then proving the contrapositive of (10): "If a triangle is not equiangular, then it is not equilateral." Or, we might prove both (10) and (11), or the contrapositives of both (10) and (11).

EXERCISES 3.1

1. State the converse and contrapositive of each of the following.
    a) If two triangles are similar, then the angles of one are congruent to the angles of the other.
    b) If 5 is larger than 6, then —1 is larger than 0.
    c) When the weather is bad, I stay inside.
    d) I get nervous whenever I take an examination.
    e) I will go only if you go.
    f) He wears a coat only when the temperature is below freezing.
    g) Your manner of doing things implies that you are stupid.
    h) Your excellent diction indicates that you are well educated.
    i) All men are mortal.
    j) Every person I've spoken to is going.

2. Prove that $A \rightarrow B$ and $\sim A \rightarrow \sim B$ are not logically equivalent statements. To what statement is $\sim A \rightarrow \sim B$ logically equivalent?

3. Determine whether or not the two statements in each of the following are logically equivalent.
    a) If I go, you will go. If you do not go, I will not go.
    b) I will go only if you go. I will not go only if you do not go.

c) When it rains, it pours. When it does not rain, it does not pour.

d) All men are greedy. No man is not greedy.

e) All equilateral triangles are isosceles. No equilateral triangle is not isosceles.

f) I will pay you only when you finish the work. If you do not finish the work, I will not pay you.

4. Prove that the contrapositive of the contrapositive of $A \rightarrow B$ is logically equivalent to $A \rightarrow B$.

3.2 ON THE VALIDITY OF A PROOF

Thus far our discussion has centered about the structure and truth value of statements. Any discussion or proof, mathematical or otherwise, consists of many interrelated statements. We usually begin with certain statements that are accepted as being true and then argue to a conclusion which we wish to prove. Any statements used to prove some conclusion are known as *premises*. Thus, a premise might be an axiom, hypothesis, definition, or previously proved statement.

As we pointed out in Chapter 1, the only type of truth with which the mathematician is concerned is essentially a relative truth. A conclusion is true or false only in relation to some axiomatic system; the conclusion is true if it follows in a logically valid manner from the axioms of the system and is false if its negation is true. Having investigated to some extent the properties of individual statements, we now turn our attention to the structure comprising several statements; in particular, we wish to discuss the question: When is a proof logically valid?

What is it that we really want from a chain of reasoning? We want to be certain that if all the premises used in the chain of reasoning are true, then the conclusion must necessarily also be true.

Consider the following argument.

12. Premise 1: I am either right or wrong.

Premise 2: I am not wrong.

Conclusion: I am right.

If (12) is to be a valid argument, then whenever the two premises are true, the conclusion must also be true. Let A be "I am right" and B be "I am wrong." Then (12) becomes

13. Premise 1: $A \vee B$

Premise 2: $\sim B$

Conclusion: A

Now we can consider the truth table 3.2. The important thing to note is that when $A \lor B$ and $\sim B$ have the value "true" (row 1), A also is true. Thus, whenever the premises in (12) are true, the conclusion must also be true; therefore (12) is a valid argument.

Table 3.2

A	$\sim B$	B	$A \lor B$
T	T	F	T
T	F	T	T
F	T	F	F
F	F	T	T

The reader must understand that we are not saying that an argument is valid if the conclusion is in fact true. The touchstone of a valid argument is that the conclusion is true if every premise used in the argument is true. It is quite possible to arrive at a factually false conclusion by means of a logically valid argument if one of the premises happens to be factually false. Example 3 illustrates this point.

Example 3

Premise 1: Either I am a cat or I am a dog.

Premise 2: I am not a cat.

Conclusion: I am a dog.

The argument here is of the same form as (13), hence is valid. But premise 1 is factually false. If premises 1 and 2 were both true, in fact, then the conclusion would also have to be true in fact.

Example 4. We see from Table 3.2 that the following argument is not valid.

Premise 1: $A \lor B$

Premise 2: A

Conclusion: $\sim B$

We can see that if $A \lor B$ is true and A is true, then $\sim B$ may be either true or false (rows 1 and 2).

Thus, to answer the question, "Is a particular chain of reasoning valid?" we must be able to answer this question: "When all the premises in the chain of reasoning are true, is the conclusion necessarily true?" As a rule, the best way to find out is by means of a truth table.

Example 5. We will investigate the validity of the following argument.

Premise 1: All men are mortal.

Premise 2: All mortals need water.

Conclusion: All men need water.

Table 3.3

A	B	C	$A \to B$	$B \to C$	$A \to C$
T	T	T	T	T	T
T	T	F	T	F	F
T	F	T	F	T	T
T	F	F	F	T	F
F	T	T	T	T	T
F	T	F	T	F	T
F	F	T	T	T	T
F	F	F	T	T	T

Premise 1 can be rephrased as "If an object is a man, it is mortal" and premise 2 can be rephrased as "If an object is mortal, it needs water." If we let A be "The object is a man," B be "The object is mortal," and C be "The object needs water," then the argument becomes

Premise 1: $A \to B$

Premise 2: $B \to C$

Conclusion: $A \to C$

The appropriate truth table for determining the validity of the argument is Table 3.3. Note that once we find the truth values of A, B, and C, we can get the truth values of $A \to B$, $B \to C$ and $A \to C$ from Table 2.10. The critical observation is that whenever premises 1 and 2 are true (rows 1, 5, 7, and 8), the conclusion is also true. Therefore the argument is valid.

EXERCISES 3.2

1. Verify Table 3.3.
2. Use Table 3.3 to show that the following is not a valid argument.

   Premise 1: If it is snowing, it is cold.

   Premise 2: If it is snowing, I will wear a coat.

   Conclusion: If it is cold, I will wear a coat.

3. Prove that each of the following is a valid argument.

   a) $A \to B$
   $$\frac{A}{B}$$

   b) $A \to B$
   $$\frac{\sim B}{\sim A}$$

   c) A
   B
   $$\overline{A \wedge B}$$

   d) $(A \to B) \wedge (C \to D)$
   $$\frac{A \vee C}{B \vee D}$$

4. Prove that each of the following is an invalid argument.

a) $A \rightarrow B$
    $\sim A$
    _____
    $\sim B$

b) $A \rightarrow B$
    B
    _____
    A

3.3 SOME ELEMENTARY ARGUMENTS

In this and in Sections 3.4 and 3.5, we shall look at a number of simple valid argument forms as well as some invalid argument forms which have the appearance of being valid.

Two of the most simple and classical argument structures are *modus ponens* and *modus tollens*. The form of a *modus ponens* argument is as follows.

14. *Modus Ponens:*

   Premise 1: $A \rightarrow B$

   Premise 2: A
   _____

   Conclusion: B

Example 6

   Premise 1: If Socrates is a man, he is mortal.

   Premise 2: Socrates is a man.
   _____

   Conclusion: He is mortal.

Row 1 of Table 2.10 shows the validity of the modus ponens argument form. We were asked to prove it in Exercise 3(a) of the last exercise set. A statement of the modus ponens form is: If both premises $A \rightarrow B$ and A are true, then the conclusion B must also be true. We may also rephrase this argument as: If A occurs, B occurs; A occurs, hence B occurs.

The form of a *modus tollens* argument is given by the following.

15. *Modus Tollens:*

   Premise 1: $A \rightarrow B$

   Premise 2: $\sim B$
   _____

   Conclusion: $\sim A$

Example 7

   Premise 1: If it is raining, then I am wearing my coat.

   Premise 2: I am not wearing my coat.
   _____

   Conclusion: It is not raining.

We were asked to prove the validity of the modus tollens argument in Exercise 3(b), Section 3.2.

We may rephrase the modus tollens argument as: If A occurs, B occurs. But B is not occurring, hence A is not occurring.

In the statement $A \rightarrow B$, A is called the *antecedent* and B is the *consequent*. Resembling the valid argument forms of modus ponens and modus tollens are two *invalid* argument forms called the *Fallacy of Affirming the Consequent* and the *Fallacy of Denying the Antecedent*. These invalid arguments are as follows.

16. *Fallacy of Affirming the Consequent:*

Premise 1: $A \rightarrow B$

Premise 2: B

Conclusion: A

17. *Fallacy of Denying the Antecedent:*

Premise 1: $A \rightarrow B$

Premise 2: $\sim A$

Conclusion: $\sim B$

The reader was asked to prove the invalidity of these argument forms in Exercise 4, Section 3.2. Now we shall review the reason for the invalidity of (16). Looking at Table 2.10, we see that if the premises $A \rightarrow B$ and B in (16) are true, it is possible for A to be either true or false. Thus the truth of the conclusion does not follow necessarily from the truth of the premises. Argument (16) is therefore invalid.

Now let us consider several fallacious arguments which are of the form of (16) and (17).

Example 8. An example of affirming the consequent is:

Premise 1: If you are smart, then I am smart.

Premise 2: I am smart.

Conclusion: You are smart.

Even if both premises are true, it is quite possible for me to be smart without you being smart. Your being smart is a sufficient condition for my being smart, but not a necessary one. In fact, it may be that "You are smart" is true. If the premises in an invalid argument are true, then the conclusion need not be false. The argument is invalid only because the truth of the premises does not *force* the conclusion to be true.

Another example of affirming the consequent is given by the following.

Example 9

Premise 1: If Socrates is a man, he is mortal.

Premise 2: He is mortal.

Conclusion: Socrates is a man.

Compare this with the valid argument using the same statements in Example 6.

Example 10. Now let us consider some instances of denying the consequent.

a) Premise 1: If it is snowing, it is cold.

Premise 2: It is not snowing.

Conclusion: It is not cold.

b) Premise 1: If Sue is a boy, she is a human being.

Premise 2: Sue is not a boy.

Conclusion: She is not a human being.

In (a), of course, it might be cold, even though it is not snowing. Premise 1 states that snow is sufficient for cold, but not necessary. Analogous objections apply to argument (b).

The last valid argument form we will consider in this section is the *Hypothetical Syllogism*. The form for this argument is as follows:

18. *Hypothetical Syllogism:*

Premise 1: $A \rightarrow B$

Premise 2: $B \rightarrow C$

Conclusion: $A \rightarrow C$

Example 11

Premise 1: If you are right, I am wrong.

Premise 2: If I am wrong, I am sorry.

Conclusion: If you are right, I am sorry.

The validity of the hypothetical syllogism was shown in Example 5. Another illustration of the hypothetical syllogism is the following example.

Example 12

    Premise 1: All men can walk (i.e., if an object is a man, it can walk).

    Premise 2: Anything that can walk has legs (i.e., if an object can walk, it has legs).

    Conclusion: All men have legs (i.e., if an object is a man, it has legs).

Resembling the hypothetical syllogism, but invalid, is the following argument form.

19. Premise 1: $A \rightarrow B$

    Premise 2: $A \rightarrow C$

    Conclusion: $B \rightarrow C$

The reader was asked to prove the invalidity of this argument in Exercise 2, Section 3.2. For another illustration of this fallacious argument, let us consider the following example.

Example 13

    Premise 1: All tigers are animals.

    Premise 2: All tigers have stripes.

    Conclusion: All animals have stripes.

Another invalid argument that is similar to the valid hypothetical syllogism is

20. Premise 1: $B \rightarrow A$

    Premise 2: $C \rightarrow A$

    Conclusion: $B \rightarrow C$

Example 14. The following illustrates argument (20).

    Premise 1: All men are animals.

    Premise 2: All tigers are animals.

    Conclusion: All men are tigers.

EXERCISES 3.3

1. Identify the type of argument used in each of the following structures and indicate whether the argument is valid or invalid.

    a) If it is raining, then I am not going.

       It is raining.

       I am not going.

b) If Jack is smart, he will leave.

If Jack leaves, he will go home.

If Jack is smart, he will go home.

c) If I am a professor, I am smart.

I am smart.

I am a professor.

d) If you read the book, you would know the capital of India.

You do not know the capital of India.

You did not read the book.

e) I eat when I am hungry.

I am eating.

I am hungry.

f) If all men were good, there would be no wars.

Some men are not good.

There will be wars.

g) All professors can read.

All professors can think.

All who can read can think.

h) If it is cold, I will stay home.

If it is cold, I will drink hot chocolate.

If I stay home, I will drink hot chocolate.

i) No man is an elephant (i.e., elephant → ~man).

No elephant is President.

No man is President.

j) I am singing.

When I am singing, I sound awful.

I sound awful.

2. In each of the following, two premises are given. Indicate any valid conclusion that may be drawn from these premises and the argument used to arrive at that conclusion. If no conclusion can be drawn, write "none."

a) If I stay here, I will be in danger. But I will stay here anyway.

b) All men want to be rich. However, all who want to be rich are unhappy.

c) If I were king, I would be happy. I am not happy.

d) If I were king, I would be powerful. But I am not king.

e) If I were king, I would live in a castle. I do live in a castle.

f) I will come only if you call. You are calling.
g) I eat only when I am hungry. I am not hungry.
h) All lions eat meat. All lions have sharp teeth.
i) No lion eats grass. Any animal that eats grass is a cow.

3. Show that argument (20) is an invalid argument.

3.4 MORE ELEMENTARY ARGUMENTS

Another valid argument form is the so-called *Disjunctive Syllogism.* This
argument runs as follows.

21. *Disjunctive Syllogism:*

Premise 1: $A \lor B$

Premise 2: $\sim B$

Conclusion: A

This argument was considered at some length in Section 3.2 and was
shown then to be valid. Another specific example of (21) is given below.

Example 15

Premise 1: It is either dangerous or dull.

Premise 2: It is not dangerous.

Conclusion: It is dull.

If it is possible for both A and B to be true simultaneously, then the
following is an invalid argument.

22. Premise 1: $A \lor B$

Premise 2: B

Conclusion: $\sim A$

On the other hand, if A and B cannot be true simultaneously (for example,
if B is the negation of A) then (22) is a valid argument. Let us consider
a valid example and an invalid example which both use (22).

Example 16

Valid:

Premise 1: The water is hot or cold.

Premise 2: The water is hot.

Conclusion: The water is not cold.

Invalid:

Premise 1: This object is red or a pencil.

Premise 2: This object is red.
_____

Conclusion: This object is not a pencil.

Two other valid argument forms are the *Constructive Dilemma* and the *Destructive Dilemma*. We will look at the form of each, together with an example.

23. *Constructive Dilemma:*

Premise 1: $(A \rightarrow B) \wedge (C \rightarrow D)$

Premise 2: $A \vee C$
_____

Conclusion: $B \vee D$

Example 17

Premise 1: If it snows I will go sledding; but if it is warm, I will go for a walk.

Premise 2: It will either snow or be warm.
_____

Conclusion: I will go sledding or go for a walk.

Note that the constructive dilemma is essentially a combination of two modus ponens arguments.

24. *Destructive Dilemma:*

Premise 1: $(A \rightarrow B) \wedge (C \rightarrow D)$

Premise 2: $\sim B \vee \sim D$
_____

Conclusion: $\sim A \vee \sim C$

Example 18

Premise 1: If I speak, I will be hanged; and if I am silent, I will be shot.

Premise 2: I will not be shot, or I will not be hanged.
_____

Conclusion: Either I will not speak, or I will not be silent.

The destructive dilemma is essentially the combination of two modus tollens arguments.

To conclude this section, we present two arguments, each of which claims to prove "The butler did it." We will analyze each argument step by step to determine whether or not it is logically valid.

Example 19. Either the butler was in his room that night, or he was in the library with Mr. van Smyth. If the butler was in his room, then the light in his room was on. There was no light on in the butler's room, hence the butler was in the library. Either the butler was alone with Mr. van Smyth or someone else was also present. If someone else was also present, the butler did not do it. But if the butler was alone with Mr. van Smyth, then no cry for help would have been heard. No cry for help was heard. Therefore the butler was alone with Mr. van Smyth. Consequently, it was the butler who did it.

This argument can be analyzed as follows.

25. If the butler was in his room, then the light in his room was on.

The light in his room was not on.

Conclusion: The butler was not in his room.

The conclusion of (25) is a valid conclusion, arrived at by using modus tollens. Continuing, we analyze:

26. Either the butler was in his room, or he was in the library.

The butler was not in his room [conclusion from (25)].

Conclusion: The butler was in the library.

Argument (26) is a valid argument using the disjunctive syllogism structure.

27. If the butler was alone with Mr. van Smyth, then no cry for help would have been heard.

No cry for help was heard.

Conclusion: The butler was alone with Mr. van Smyth.

Argument (27) is invalid because it uses the fallacy of affirming the consequent. In this case, we have not excluded the possibility that no cry for help would have been heard even with others present. Even though the conclusion for (27) was arrived at invalidly, let us accept it for the sake of completing this example.

28. If the butler was not alone, he did not do it.

The butler was alone [invalid conclusion from (27)].

Conclusion: The butler did it.

Argument (28) is an invalid argument because it applies the fallacy of denying the antecedent. Thus, even if the butler did do it, we have not logically proved our case since not one, but two, invalid arguments were used in the course of the proof.

Example 20. If it was warm, Mr. van Smyth would have been on the patio; but if the butler had called him to the phone, he would have been in the library. Mr. van Smyth was either not in the library or not on the patio. Hence either it was not warm, or the butler did not call him. It was warm; therefore the butler must not have called him. But if the butler did not call him, then the butler was not in his room. And if the butler was not in his room, then the butler did it. Therefore the butler did do it.

This argument is analyzed as follows:

29. If it was warm, Mr. van Smyth would have been on the patio; but if the butler called him to the phone, he would have been in the library.

Mr. van Smyth was either not in the library or not on the patio.

Conclusion: Either it was not warm, or the butler did not call him.

Use of the destructive dilemma makes (29) a valid argument.

30. Either it was not warm, or the butler did not call him.

It was warm.

Conclusion: The butler did not call him.

Argument (30) is a valid argument because it uses disjunctive syllogism.

31. If the butler did not call him, then the butler was not in his room.

The butler did not call him.

Conclusion: The butler was not in his room.

Argument (31) is a valid modus ponens argument, as is the final argument for Example 20, by which we prove validly that the butler did do it.

If the butler was not in his room, then the butler did it.

The butler was not in his room.

Conclusion: The butler did it.

EXERCISES 3.4

1. Determine whether each of the following arguments is valid or invalid.

 a) Either I am right or I am wrong.

 I am not wrong.

 I am right.

 b) Either that pan is black or made of iron.

 That pan is not black.

 That pan is made of iron.

c) He went hiking or fishing.

   He went hiking.
   _____

   He did not go fishing.

d) You will hate me if I don't invite you and Sue will hate me if I do.

   I will invite you, or I will not invite you.
   _____

   Either Sue will hate me, or you will hate me.

e) If I go to Ted's, I will see Sally, but if I go to Sam's, I will see Sue.

   I will see either Sally or Sue.
   _____

   I will go to Ted's, or I will go to Sam's.

f) I get indigestion when I eat pizza and have nightmares when I drink beer.

   I have neither indigestion nor nightmares.
   _____

   I have neither eaten pizza nor drunk beer.

g) If I am sad, I watch a sunset, and if I am happy, I read a book.

   I am not watching a sunset, but neither am I happy.
   _____

   I am not sad and I am not reading a book.

2. In each of the following, two premises are given. Indicate any valid conclusion that we can draw from these premises and the argument used to arrive at that conclusion. If no conclusion can validly be drawn, write "none."

   a) I am either wise or foolish. But I am certainly not foolish.

   b) Either I wrote the play, or I am directing the play. I wrote the play.

   c) When I see you cry, it makes me sad, but when I see you laugh, I am happy. I always see you either crying or laughing.

   d) When you want something cheap, you go to Grubby's, but when you want something good, you go to Goodie's. Either you are not going to Grubby's or you are not going to Goodie's.

   e) I will not talk, neither will I run away. But if I talk, I will be shot, and if I run away, they will find me.

   f) When I run, I get tired if I don't run slowly. I am not tired.

3. Note that in part (d) of Exercise 1 the second premise, "I will invite you, or I will not invite you," is a tautology (Section 2.6). Is it ever necessary to state a tautology explicitly as a premise in an argument, i.e., is there ever any possibility that a tautology will be false? Suppose that some conclusion is drawn in a logically valid manner from two premises and that one of the premises is a tautology. Explain why the truth of the conclusion follows from the truth of that premise which is not a tautology.

4. Prove that (22) is a valid argument if A and B cannot both be true simultaneously. [*Hint:* Make a truth table; for the row in which A and B are both true, make "$A \lor B$" false.]

3.5 PROOFS BY CONTRADICTION

A particular bête noire of mathematical novices, and even of some advanced students of mathematics, is the *proof by contradiction*, or *reductio ad absurdum*.

Let us recall that an argument is logically valid if whenever its premises are true its conclusion is also true (Section 3.2). Thus, if the premises in a logically valid argument lead to a conclusion known to be false, then at least one of the premises must be false. In particular, if we have a logically valid argument with one premise whose truth has already been established and one premise whose truth or falsity is uncertain, and if we prove by the argument a conclusion that is known to be false, then the premise in question must be false.

Example 21. Suppose we are not sure whether the statement

32. The sum of the interior angles of a triangle is less than two right angles

is true or false in the context of Euclidean plane geometry. Statement (32), together with certain other propositions known to be true in plane geometry, can be used to prove in a logically valid manner

33. From a point P outside a line L, there is no line through P parallel to L.

If we have already proved, or assumed, that (33) is false for Euclidean plane geometry, then (32) must be false.

Every proof that some statement is false, however, is equivalent to a proof that the negation of that statement is true. For example, in proving that (32) is false, we have actually proved that any negation of (32), for example,

34. The sum of the interior angles of a triangle is not less than two right angles, i.e., the sum is at least two right angles,

is true.

In summary, we may prove that some statement is true by showing that its negation is false. We can show that the negation is false by showing that the negation, when used with premises known to be true, leads through a logically valid argument to a conclusion known to be false. This technique is called *proof by contradiction*.

Example 22. Let us accept the following premises as being true.

35. All men are mortal.

36. Socrates is a man.

We use a proof by contradiction to prove

37. Socrates is mortal.

The negation of (37) is

38. Socrates is not mortal.

If (38) is true, then the statement

39. Some (i.e., at least one) men are not mortal

is true. But (39) is the negation of (35). Since (35) is true, (39) must be false. Since (39) is false, (38) is false. Therefore (37) is true.*

We now give another example of a proof by contradiction.

Example 23. We accept the following premises as true.

40. Any time it is snowing, it is cold.

41. It is snowing.

We wish to prove

42. It is cold.

Of course (42) follows from (40) and (41) directly by modus ponens, but we will carry out a proof now by contradiction.

If it is not cold, i.e., if the negation of (42) is true, then we have

43. There is a time when it is snowing, but not cold.

However, (43) is the negation of (40). Since (40) is true, (43) is false. Therefore "It is not cold" is false. Hence (42) is true.

In any proof by contradiction certain precautions must be observed.

If we wish to prove that some statement A is true by showing that the negation of A leads to a false conclusion, we must be certain that *we use a genuine negation* of A. For example, by showing that (32) is false, we proved that (34) is true. If, however, we wished to prove

44. The sum of the interior angles of a triangle is equal to two right angles,

then proving that (32) is false would be insufficient since (32) is not the negation of (44).

---

* That is: $(38) \equiv \sim(37)$. $(38) \rightarrow (39)$. $(39) \rightarrow \sim(35)$. But $\sim(35)$ is false, hence (39) is false. Therefore (38) is false. But then $\sim(38) \equiv (37)$ is true.

We must be certain that, in showing that the negation of a statement we wish to prove leads to a false conclusion, *we use only premises known to be true as well as a logically valid argument.* For if the negation of our statement is not the only statement whose truth value is uncertain, or if the argument used is not logically valid, then we could arrive at a false conclusion even if the negation of our statement happened to be true.

We must be certain that *we arrive at a genuinely false conclusion.* The great eighteenth-century geometer Girolamo Saccheri felt that he had proved Euclid's famous parallel postulate because he found that by using the negation of the parallel postulate, he could show that the sum of the interior angles of a triangle was not two right angles. In fact, however, such a finding is a perfectly valid conclusion in the type of non-Euclidean geometry with which Saccheri was unknowingly working.

It should also be noted that many statements which can be proved by contradiction can also be proved directly by other means, e.g., see Example 23. It then becomes a matter of taste as to which proof is preferable, that is, more esthetically pleasing, simpler, or more elegant.

The proof by contradiction may be summarized in notation as follows.

45. To prove A:

$A \lor \sim A$	tautology
$\sim A \to B$	$\sim A$ used to prove B
$\sim B$	B is false
$\sim(\sim A)$	modus tollens
A	disjunctive syllogism, or $A \equiv \sim(\sim A)$

EXERCISES 3.5

1. Analyze each of the following arguments to determine whether they are proofs by contradiction. Determine in each case whether the conclusion has been arrived at validly.

    a) Either Sam will be elected, or Sam will not be elected. If Sam is not elected, it will be because he is not running. But Sam is running. Therefore Sam will be elected.

    b) If all triangles are isosceles, then there is no triangle which has three noncongruent angles. There is, however, a triangle with three noncongruent angles. Therefore all triangles are not isosceles. [The reader should note that not all the steps in (45) need be stated explicitly.]

    c) Either it is raining, or it is cool. If it is cool, I will wear a coat. I am not wearing a coat; hence it is not cool. Therefore it is raining.

    d) If 4 is less than 1, then 3 is less than 0. But 3 is not less than 0, hence 4 is not less than 1.

2. Prove each of the following statements by contradiction (even though you might be able to prove each statement directly).

   a) If N is any positive integer, then there is an even integer greater than N.
   b) All horses are animals.
   c) If an integer k is odd, then $k + 1$ is even.

3. The following argument forms bear some resemblance to (45). Examine each to determine whether it is logically valid. If it is valid, indicate which arguments it combines.

   a) $A \lor B$
      $A \to C$
      $\sim C$
      $\underline{\sim A}$
      B

   In what case would this be a genuine proof by contradiction?

   b) $A \lor \sim A$
      $C \to A$
      $\sim C$
      $\underline{\sim A}$
      A

   c) $A \lor B$
      $\sim (A \land B)$
      $A \to C$
      C
      $\underline{A}$
      $\sim B$

3.6 SUMMARY

We summarize below all the valid arguments presented in this chapter with an illustration of each.

Modus Ponens:

   $A \to B$
   $\underline{A}$
   B

Illustration:

   If the weather is bad, the party is canceled.

   The weather is bad.

   The party is canceled.

Modus Tollens:

$$A \rightarrow B$$
$$\sim B$$
$$\overline{}$$
$$\sim A$$

Illustration:

If you are good, you will get a cookie.

You will not get a cookie.

You are not good.

Hypothetical Syllogism:

$$A \rightarrow B$$
$$B \rightarrow C$$
$$A \rightarrow C$$

Illustration:

If I feel better, I will come to the party.

If I come to the party, I will see Sally.

If I feel better, I will see Sally.

Disjunctive Syllogism:

$$A \vee B$$
$$\sim B$$
$$A$$

Illustration:

I will go skating or skiing.

I will not go skating.

I will go skiing.

Constructive Dilemma:

$$(A \rightarrow B) \wedge (C \rightarrow D)$$
$$A \vee C$$
$$B \vee D$$

Illustration:

If I see Jack, I will apologize; but if I see Sam, I will hit him.

I will either see Jack or Sam.

I will either apologize to Jack, or hit Sam.

Destructive Dilemma:

$(A \to B) \land (C \to D)$

$\underline{\sim B \lor \sim D}$

$\sim A \lor \sim C$

Illustration:

If I play cards, I rest; and if I play golf, I exercise.

$\underline{\text{Either I will not rest, or I will not exercise.}}$

Either I will not play cards, or I will not play golf.

Proof by Contradiction:

$(A \lor \sim A)$

$\sim A \to B$

$\sim B$

$\underline{\sim(\sim A)}$

A

Illustration (Any modus tollens example is essentially a proof by contradiction. As another example we give the following.): Suppose a statement C is true if and only if a statement D is true. Then C is false if and only if D is false. For if C is false and D is true, then we could not say that C is true if D is true. And if C were true when D was false, then we could not say that C is true only if D is true.

EXERCISES 3.6

Each of the following arguments is to be analyzed in the manner of Examples 19 and 20.

1. Any plane triangle is either isosceles, or not. If any triangle is not isosceles, then there is a triangle with no two angles congruent. But there is a triangle with no two angles congruent. Therefore any triangle is not isosceles.

2. If Sam runs, he will be elected. If he is elected, either he will make a good president, or he will be unhappy. If the students do not cooperate, then Sam will not make a good president. The students will not cooperate. Therefore if Sam runs, he will be unhappy.

3. All men want power. But if anyone wants power, he will stop at nothing. If all men will stop at nothing, then there will be wars. And if there will be wars, then people will be killed. Therefore people will be killed.

4. When Sam wants anything, he will either buy it or steal it. If he steals it, he will be sent to jail. If he is sent to jail, he will not be able to finish college.

Sam will be able to finish college. Therefore if Sam wants anything, he will buy it.

5. If I go to the party, I shall see Sue; and if I see Sue, she will not speak to me. If I do not go to the party, then I will have to stay home. However, I will not be happy if I stay home. I must be happy, or I will not be able to do my homework this weekend. Therefore I will not be able to do my homework this weekend.

6. If I go to the movie and if I like it, I will tell you about it. But if I do not go to the movie, I will watch television and play cards. If I watch television, I will be bored; and if I play cards, I will lose money. I cannot afford to lose any money, so I must go to the movie. But I am sure I will not like the movie, hence I am sure I will not tell you about it.

7. If my opponent is right, you should not vote for me. If he is right, then I am either a fool or a scoundrel. If I were a fool, I would not be a senator; and if I were a scoundrel, I would be in jail. But I am a senator and I am not in jail. Therefore you should vote for me.

8. If the apparatus was defective, then the experiment would not have worked. If the experiment had not worked, then we would not have made the discovery. If we had not made the discovery, our competitors would have found the process first. If our competitors had found the process first, then they would have sold it either to Russia or France. If they had sold it to Russia, we would have been lost; but if they had sold it to France, France would have kept it secret. Our competitors would not have sold it to Russia. Therefore if the apparatus was defective, France would have kept it secret.

9. Assume n^2 is even, but n is odd; then $n = 2k + 1$, where k is some integer. But then

$$n^2 = (2k + 1)^2 = 4k^2 + 4k + 1 = 2(2k^2 + 2k) + 1.$$

Thus n^2 is odd, contradicting the assumption that n^2 is even. Therefore if n^2 is even, n is even.

Sets

4.1 THE NOTION OF A SET

In this chapter we shall consider collections of objects only as collections of objects. For example, if we discuss a herd of cows in a field, we see the herd only as a group of objects; the fact that these objects are cows or that they are grazing in the field is quite immaterial.

The importance of this study should be clear, even to students with little background in mathematics. For everything studied by mathematicians, or by anyone for that matter, is a collection of objects of some type. The reader may object that in practice we never consider a collection of objects without caring what those objects are. This is true; but understanding the concept of a collection of objects* surely serves as a foundation for the study of a collection when we do care what objects are in the collection.

If we compare developing a mathematical system to building a house, then the axioms are the foundation on which the house rests and the propositions which we derive from the axioms are the material of which most of the house is built. Logic is the system of natural laws which govern how the house must be built, the principles of its construction, if you prefer. But the theory of sets is the earth in which the foundation is dug; earth which also serves as the source of the raw material from which much of the house is produced. Set theory underlies all mathematical systems.

The collections we shall consider in this chapter may consist of any objects whatsoever, either real or imagined. A collection may consist of the pieces of furniture in a house, three red and three green marbles, the cows in a field, or perhaps a collection of three cows and two green marbles. Not all objects in a collection need be of the same type. We shall make the

---

* That is, considering a collection of objects as an abstraction.

one stipulation, however, that any collection we consider must be *well defined*. Given any object whatever, we must be able to decide, at least in theory, whether or not that object is in the collection.*

For example, if the collection consisted solely of our brothers and sisters, then we would know each member of the collection personally and would be able to identify objects which belonged or did not belong to the collection. For most other collections, we must have a rule for deciding what belongs to the collection. For example, a collection might consist of all brown dogs. There are obviously more brown dogs than anyone could ever know or keep in mind, but if we were presented with any object, we could tell whether or not that one particular object was or was not a brown dog. Thus we could effectively decide which objects are in and which objects are out of the collection of all brown dogs.

Example 1. We have seen that the collection of brown dogs is well defined since we can determine whether any given object is a brown dog. Similarly, the collection of cats whose owners live in New York City is a well defined collection. The collection of excellent novels, however, is not well defined. For if we are given any book, we might be able to determine whether or not it is a novel, but whether or not it is excellent is purely a matter of opinion. No amount of knowledge on our part could ever really resolve the question of whether certain novels are excellent.

We begin to see from Example 1 why we want the collections we consider to be well defined. Even if we are considering a collection of objects only as a collection of objects, we want the collection to be unambiguous. Thus, if we were considering a herd of cows just as an abstract collection, we would be somewhat disconcerted if cows were added to and subtracted from the herd while we were making the study, or if someone threw in a few butterflies after we had begun our investigation. While mathematicians may deal with the abstract, it is the precise abstract. Ambiguity and subjectivity have no place in any truly mathematical discussion.

Example 2. Consider the following well-known paradox: The barber in East Grunion, Ohio, shaves a male of East Grunion if and only if that male does not shave himself. Consider the collection of all males in East

---

* We add "at least in theory" because there are some collections for which there is as yet no practical way to decide whether certain objects are in the collection. For example, it is not known whether certain constants encountered in certain mathematical studies are rational or irrational. But this indetermination is due to a lack of knowledge rather than to any inherent logical difficulty; in other words, we could decide whether the number was rational or irrational if we knew enough about it.

Grunion who shave themselves. Is the barber a member of this collection? If the barber is a member of this collection, then he shaves himself. But then, according to our criterion, he is not a member of the collection of men whom he shaves, i.e., if the barber is a man who shaves himself, then the barber does not shave himself.

Example 2 illustrates the difficulties we can get into if we do not pay careful attention to the way we say things. For if we had said that the barber shaves any *other* male in East Grunion if and only if that male does not shave himself, we would not have arrived at any contradiction.

Definition 4.1. *A set is any well-defined collection of objects. That is, a set is any collection which satisfies the requirement that, given any object whatever, we can determine whether or not that object is in the collection (provided, of course, that we are also given sufficient information about the object). The objects in a set are called elements, or points, of the set.**

Example 3. The collection of all red buildings forms a set since, given any object, we are able to determine whether or not that object is a red building. Each red building is an element of the set of all red buildings.

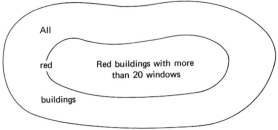

Fig. 4.1

Example 4. The collection of all red buildings that have more than 20 windows is also a set since we can determine whether any object is a red building, and we can then count the windows to see whether there are more than 20. We should note that each element of the set of red buildings with more than 20 windows is also an element of the set of red buildings. (See Fig. 4.1.) It is not necessarily true, of course, that every red building has more than 20 windows.

---

* If a collection of objects is not a set, or if it is not known whether the collection is a set, then mathematicians generally call such a collection a *class*. At the level of this text virtually all statements made about sets are equally applicable to classes. For example, we can take the union or intersection of two classes just as we would take the union or intersection of two sets.

Example 5. The collection which contains only the largest even integer is a set since, given any object, we can determine whether it is an integer, is even, and is larger than any other even integer. But note that this set contains no elements because there is no largest even integer; that is, for every even integer n, there is a still larger even integer, $n + 2$.

Example 6. The collection which contains the world's most beautiful lakes is not a set because whether or not a lake is "most beautiful" is a matter of opinion and not an objectively ascertainable property. Even the collection of the world's tallest buildings would not be a set. For although we could measure the height of any building accurately, and even determine *the* tallest building in the world, the definition of the set does not indicate how many of the world's tallest buildings we should consider, e.g., is the tenth tallest building in the world to be considered one of the world's tallest buildings, or do we want only the first five?

The reader may have wondered: If sets are the object of mathematical study, why are they not defined by means of axioms? That is, why are there no rules that define a set as something that satisfies certain axioms containing undefined terms—the approach which, in Chapter 1, was shown to be so essential to mathematics? Such an axiomatic approach to the theory of sets is possible and is used in more sophisticated study of set theory, but such an approach would not be appropriate for this book. We are therefore using a more "concrete" definition to talk about sets so that we may discuss sets more intuitively, and, it is hoped, with more advantage to the student. Since a great deal has been said about the axiomatic character of mathematics, we are including an appendix which will give not only axioms for a theory of sets, but also axioms for a "logic" as well.

EXERCISES 4.1

1. Decide whether each of the following is a set:
   a) The collection of all citizens of the United States
   b) The collection consisting of the three tallest buildings in Canada
   c) The collection containing the three greatest works of Beethoven
   d) The collection of letters of the Greek alphabet
   e) The collection containing the tallest building closest to Omaha
   f) The collection consisting of all boats which have crossed the Atlantic Ocean in less than four days

2. Are "the tallest building in New York City" and "the collection which consists of the tallest building in New York City" the same? Explain in some detail.

3. Is the collection of true statements a set? The reader should first observe that there are many statements whose truth or falsity is a matter of conjecture, for example, "There are living creatures on Mars." Thus, from this point of view, it is impossible to tell whether many statements are in the collection or not. The question may also be raised as to how to classify future events which may or may not occur, for example, "It will rain tomorrow." The reader should also try to decide whether the statement "This statement is false" belongs to the collection.

4. Consider the collection T of all sets which contain themselves as elements, that is, a set S would be a member of this collection if S itself were also a member of S. We might be tempted to say that T is a set. Now consider the collection W of all sets which do not contain themselves as elements. This too would seem to be a set. Show, however, that we are confronted with a paradox when we try to decide whether W itself is a member of W. If we approach set theory axiomatically, we would resolve the paradox by refusing to recognize as a set any collection which contains itself as an element.

4.2 SET-BUILDING NOTATION

Notation, that is, the use of symbols to represent objects, concepts, or phrases, is essential to mathematics. If mathematicians had to write everything in longhand rather than use concise symbols for complicated definitions and concepts, very little progress would occur in mathematics. In certain areas of mathematics, the rate of progress has increased, not after new concepts and ideas were invented, but rather after new and better notation was found for what was already known. For example, have you ever tried to multiply two Roman numerals? It is of course much more cumbersome (for example, try LI times CVI) than multiplication using the elegant Arabic notation, 51 times 106. A mere change in notation makes the computation immeasurably easier. Without notation there could be little useful mathematics.

We may write

1. the set of all brown dogs

as

2. $\{x \mid x \text{ is a brown dog}\}$.

We shall use braces to indicate a set; whatever falls within the braces defines the set. Any particular set may be indicated in one of two ways. First, as in (2), we may use a symbol to denote a typical element of the set (in this example, we use x). The symbol is followed by a vertical bar, and to the right of the bar, we give the criterion that the typical element must satisfy to be a member of the set. Thus, (2) could be read as

3. the set of all x such that x is a brown dog.

The second way to indicate a set is to list the elements of the set between the braces. This procedure is convenient only when there are few elements in the set. For example, the set containing only the first three positive integers as elements could be denoted by $\{1, 2, 3\}$.

The phrase "is an element of" is usually represented by the symbol $\in$. The symbol $\notin$ represents the phrase "is not an element of." As a rule, when a symbol represents some phrase, then that same symbol with a slash through denotes the negation of that phase. For example, since the symbol $=$ denotes the phrase "is equal to," $\neq$ represents the phrase "is not equal to."

Example 7. The set of all marbles belonging to Tommy Smith can be denoted by $\{t \mid t$ is a marble belonging to Tommy Smith$\}$. It could be denoted just as well by $\{w \mid w$ is a marble belonging to Tommy Smith$\}$. Each marble belonging to Tommy Smith is an element of this set, and any object is in this set only if it is a marble belonging to Tommy Smith; that is, $t \in \{t \mid t$ is a marble belonging to Tommy Smith$\}$ if and only if t is a marble belonging to Tommy Smith. It might be that Tommy Smith does not possess any marbles, in which case our set would contain no elements.

We must take care to distinguish any set from the elements it contains. To learn more about this distinction, let us look at the following examples.

Example 8. $\{A\}$ is the set which contains only the letter A. Note that A and $\{A\}$ are two different objects; $\{A\}$ is a collection, but A is a letter of the alphabet which is in $\{A\}$. The statement "A is an element of $\{A\}$" may be denoted by "$A \in \{A\}$."

Example 9. $\big\{1, \{1\}\big\}$ represents the set that consists of both 1 and the set that contains 1 as its only element. Hence it is quite possible to have one set as an element of another set. Since 2 is not an element of $\big\{1, \{1\}\big\}$, we may write $2 \notin \big\{1, \{1\}\big\}$. Note that it would be correct to write

$$\{1\} \in \big\{1, \{1\}\big\},$$

whereas in Example 8, it would be incorrect to write $\{A\} \in \{A\}$.

EXERCISES 4.2

1. Write out in words the meaning of each of the following.
   a) $\{z \mid z$ is a cow$\}$
   b) $\{w \mid w$ is a white house$\}$

c) $\{a, b, c\}$

d) $\{\{a\}, b, c\}$

e) Elsie $\in \{c \mid c$ is a bovine animal$\}$

f) $\{d\} \notin \{a, b, d, e\}$

g) $\{e\} \in \{\{a\}, \{b\}, \{d\}, \{e\}\}$

2. Write each of the following phrases notationally.

a) the collection of all blue houses

b) the collection consisting of the first four even positive integers

c) 4 is not a letter of the English alphabet

d) the set consisting only of the number 3 is not the same as the number 3

e) Sam is a member of both the basketball team and the football team

3. For each of the following pairs of sets, write symbolically a set which consists of all elements common to both sets. For example, for $\{x \mid x$ is a red building$\}$ and $\{y \mid y$ is a farmhouse$\}$, we might write $\{w \mid w$ is a red farmhouse$\}$.

a) $\{z \mid z$ is an integer$\}$ and $\{2, 3, \{5\}\}$

b) $\{p \mid p$ is a person$\}$ and $\{w \mid w$ is intelligent$\}$

c) $\{3, 4, 5, 6, 7\}$ and $\{4, 6, 7, 9, 10\}$

d) $\{d \mid d$ is a dog$\}$ and $\{a \mid a$ is an animal$\}$

e) $\{p \mid p$ is a pony$\}$ and $\{d \mid d$ is a duck$\}$

f) $\{t \mid t^2 = 1\}$ and $\{1, -1\}$

4. Determine which of the following statements are true and which are false. Correct each false statement so that the revised statement is true.

a) $\{1, 2, 3\} \in \{1, 2, 3, \{a, b\}\}$

b) $1 \notin \{\{w\} \mid w$ is an integer$\}$

c) $\{\{a\}\} \in \{s \mid s$ is a set which contains some set as an element$\}$

d) $\{\{\{1\}\}\}$ is a set containing only one element

4.3 SUBSETS. EQUALITY OF SETS

Having defined what we mean by a set, that is, a well-defined collection of objects, we now begin to investigate concepts which arise naturally in any discussion of sets. Given some set, we might wish to consider some part of it aside from the entire set; for example, given the set of all marbles, we may wish to consider only black marbles. The set of black marbles is what is called a *subset* of the set of all marbles.

Definition 4.2. *If S is a set, then a collection which consists entirely of elements which are also elements of S is called a subset of S. That is, a set T is a subset of a set S if every element of T is an element of S.*

The phrase *is a subset of* is frequently abbreviated by the symbol $\subset$. In keeping with the general rule, *is not a subset of* is denoted by $\not\subset$. Thus the statement

4. Every cow is an animal

may be written as

5. $\{y \mid y$ is a cow$\} \subset \{w \mid w$ is an animal$\}$.

In Example 4 we saw that the set of all red buildings with more than 20
windows is a subset of the set of all red buildings. On the other hand, the
set consisting of all farmhouses is not a subset of the set of all red buildings
even though these two sets may share some elements in common, namely,
all red farmhouses.

We now prove two basic properties of $\subset$.

*Proposition 1. Every set is a subset of itself. That is, if S is any set, then
$S \subset S$.*

Proof. If S is any set, then any element of S is an element of S. But this
is precisely the definition of $S \subset S$. (This is quite similar to Gertrude
Stein's "a rose is a rose . . . .")

*Proposition 2. If S, T, and W are any sets and $S \subset T$ and $T \subset W$, then
$S \subset W$.*

Proof. Since $S \subset T$, every element of S is also an element of T. But since
$T \subset W$, each element of T is also an element of W. Therefore each element
of S is also an element of W. Consequently, $S \subset W$. (See Fig. 4.2.)

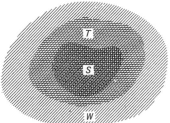

Fig. 4.2

Example 10. The set of all lame brown dogs is a subset of the set of all
brown dogs; and the set of all brown dogs is in turn a subset of the set of
all dogs. Proposition 2 (and common sense) tells us that the set of lame
brown dogs is a subset of the set of all dogs. Each of these three sets is,
of course, also a subset of itself.

We can often change a verbal argument into a set-theoretic argument
and thus establish the validity of the argument from what we know about

sets. The next chapter will be devoted to an extensive development of this idea, but one example seems appropriate at this point.

Example 11. Consider the following argument.

6. All professors are smart.

Anything that is smart is able to read.

All professors are able to read.

The reader should recall that (6) is an argument by hypothetical syllogism. From a set-theoretic point of view, (6) can be expressed as follows:

7. $\{p \mid p \text{ is a professor}\} \subset \{s \mid s \text{ is smart}\}.$

$\{s \mid s \text{ is smart}\} \subset \{r \mid r \text{ is able to read}\}.$

$\{p \mid p \text{ is a professor}\} \subset \{r \mid r \text{ is able to read}\}.$

Note that in (7) the conclusion follows from Proposition 2.

We have already encountered sets which contain no elements at all, for example, the set which consists of the largest even integer.

Definition 4.3. *The set which consists of no elements at all is called the empty set, or null set. The empty set is denoted by* $\varnothing$.

Oddly enough, while the empty set contains no elements, this very fact makes it a subset of every set. For if S is any set, then each element of $\varnothing$ (there are none) is also an element of S; hence $\varnothing \subset S$. In other words, if S is any set, but $\varnothing \not\subset S$, then there is an element x of $\varnothing$ which is not an element of S. But this means that x is an element of $\varnothing$, and consequently that $\varnothing$ contains at least one element, which is false. Therefore $\varnothing \subset S$.*

Example 12. We list now all the subsets of $\{a, b, c, d\}$: $\varnothing$, $\{a, b, c, d\}$, $\{a, b, c\}$, $\{a, b, d\}$, $\{b, c, d\}$, $\{a, c, d\}$, $\{a, b\}$, $\{a, d\}$, $\{b, c\}$, $\{b, d\}$, $\{a, c\}$ $\{c, d\}$, $\{a\}$, $\{b\}$, $\{c\}$, $\{d\}$. The original set contains only four elements, but there are $16 = 2^4$ subsets. For all sets, if a set contains n elements, it has 2^n subsets.

Note that much of this material on sets is material we have already seen in one form or another. What we are doing is merely restating these facts in a more formal manner than that in which they were previously expressed. It is a fairly common failing of the novice in mathematics to

---

* The reader should have recognized this line of reasoning to be a proof by contradiction.

try to find something complicated in the midst of simplicity; thus he misses points that in other circumstances would come easily to him. There are indeed profound questions in the theory of sets, but what has been presented so far should strike the reader as being primarily common sense. We see this point illustrated in the following discussion of the equality of sets.

Suppose Tom Smith lives at 476 Hamhocks Avenue in East Grunion, Ohio. We can see that {the house at 476 Hamhocks Avenue in East Grunion, Ohio} and {Tom Smith's house in East Grunion} are really the same set even though they are stated differently. The reason they are the same set is that they both contain precisely the same elements, namely, the house that Tom Smith lives in. Similarly,

8. $\{1, -1\}$ and   **9.** $\{t \mid t^2 = 1\}$

are also the same since they too contain exactly the same elements. More precisely, each element of (8) is an element of (9) and each element of (9) is an element of (8), that is,

$$\{1, -1\} \subset \{t \mid t^2 = 1\} \quad \text{and} \quad \{t \mid t^2 = 1\} \subset \{1, -1\}.$$

Put another way, an object is an element of (8) if and only if it is an element of (9).

To generalize these findings, we say that two sets S and T are equal if each element of S is an element of T, and each element of T is an element of S as well. That is, $S = T$ if and only if $S \subset T$ and $T \subset S$.

EXERCISES 4.3

1. Write out in good English each of the following. For example,

$$\{1, 2\} \subset \{z \mid z \text{ is an integer}\}$$

might be phrased "1 and 2 are integers."

   a) $\{t \mid t \text{ is a turtle}\} \subset \{a \mid a \text{ is an amphibian}\}$
   b) $\{\text{Tom Smith's house}\} \not\subset \{r \mid r \text{ is a red building}\}$
   c) $\{\text{Elsie, Daisy}\} \subset \{x \mid x \text{ is a cow}\}$
   d) $\{w \mid w \text{ is a whale}\} \not\subset \{f \mid f \text{ is a fish}\}$ [*Warning:* Be sure that you say no more than what you are entitled to say.]
   e) $\{n \mid n \text{ is an integer larger than 5}\} \subset \{m \mid m \text{ is an integer larger than 3}\}$
   f) $\{y \mid y \text{ is an integer which is divisible by two}\} = \{m \mid m \text{ is an even integer}\}$
   g) $\{\text{Sam, Tom, Sally}\} = \{y \mid y \text{ is a child of Mr. and Mrs. Brown}\}$

2. Write out all of the subsets of $\{1, 2, h\}$.

3. Is $\{A\} \subset S$ the same as $A \in S$? Is $\varnothing$ the same as $\{\varnothing\}$?

4. Prove the equality, or inequality, of the sets in the following pairs.

   a) $\{x \mid x$ is an odd integer$\}$ and $\{w \mid w = 2k + 1$ for some integer $k\}$
   b) $\{3, 4\}$ and $\{y \mid y^2 - 7y + 12 = 0\}$
   c) $\{t \mid t$ is an integer divisible by 2, 3 or 5$\}$ and $\{z \mid z$ is an integer$\}$

5. Indicate all relations of containment between the following sets: $\{z \mid z$ is a lame brown dog$\}$, $\{s \mid s$ is an animal$\}$, $\{$a dog named Charlie who has red fur$\}$, $\{t \mid t$ is red$\}$, $\{s \mid s$ is a living thing$\}$.

6. Prove that all elements of the empty set are red. Prove that all elements of the empty set are not red. Is there any real contradiction here?

7. Suppose $S \subset T$, $T \subset W$, and $W \subset S$. Prove that $S = W$.

8. A set S is said to be a *proper subset* of the set T if $S \subset T$, but $S \neq T$. In each of the following $\subset$ is to be interpreted as "is a *proper* subset of"; prove each of the following.

   a) If $S \subset T$ and $T \subset W$, then $S \subset W$
   b) If $S \subset T$ and $T \subset W$, then $W \not\subset S$
   c) If $S \subset T$ and $W \subset S$, then $W \subset T$

4.4 UNION AND INTERSECTION

Given two sets S and T, we may want to consider all the elements of both sets taken together, or all the elements common to both sets. For example, if we know that

$$S = \{x \mid x \text{ is a student taking some English course}\}$$

and

$$T = \{y \mid y \text{ is a student taking some history course}\},$$

then we might also wish to consider

$\{w \mid w$ is a student taking either an English course or a history course$\}$

or

$\{z \mid z$ is a student taking both an English course and a history course$\}$.

The former set is called the *union* of the sets S and T; the latter set is called the *intersection* of the sets S and T. We make the following definition.

Definition 4.4. *Let S and T be any two sets. Then the collection of objects which are elements of both S and T is called the intersection of S and T. The intersection of S and T is frequently denoted by $S \cap T$.*

The collection of objects which are elements of either S or T (or both) is called the union of S and T and is frequently denoted by $S \cup T$. Notationally,

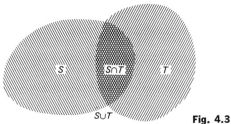

Fig. 4.3

$S \cap T = \{x \mid x \in S \ and \ x \in T\}$; $S \cup T = \{x \mid x \in S \ or \ x \in T\}$.   (*See Fig. 4.3.*)

Example 13. The intersection of $\{a, b, 3, 4\}$ and $\{4, 5, 6, 9\}$ is $\{4\}$, while the union of $\{a, b, 3, 4\}$ and $\{4, 5, 6, 9\}$ is $\{a, b, 3, 4, 5, 6, 9\}$.

Example 14. Let $S = \{r \mid r \text{ is a red object}\}$ and $T = \{b \mid b \text{ is a building}\}$. Then

$S \cup T = $ (df) $\{x \mid x \text{ is either a red object or a building (or both)}\}$

and

$S \cap T = $ (df) $\{w \mid w \text{ is both red and a building}\} = \{w \mid w \text{ is a red building}\}$.

Example 15. Let S be the set of all dogs and T be the set of all animals. Then $S \subset T$. Now $S \cup T = $ (df) $\{y \mid y \text{ is either a dog or an animal}\}$. But if an object is a dog, it is also an animal; therefore

$$S \cup T = \{y \mid y \text{ is an animal}\} = T.$$

On the other hand, $S \cap T = \{z \mid z \text{ is both a dog and an animal}\}$. Since any dog is also an animal, $S \cap T = \{z \mid z \text{ is a dog}\} = S$.

Taking the union of two sets may strike the reader as being similar to adding two numbers; and, indeed, there are quite a few similarities. The next few propositions give some of the basic properties of $\cup$; we compare these properties to the corresponding properties of $+$.

Proposition 3. *If S and T are any two sets, then $S \cup T = T \cup S$, that is, the union of S and T is equal to the union of T and S.*

Proof. $S \cup T$ consists of those objects which are either elements of S or elements of T, while $T \cup S$ consists of those objects which are either elements of T or elements of S. Clearly then, both these sets contain precisely the same elements, hence are really the same set.

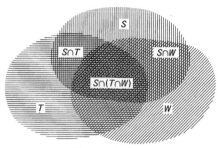

$$S \cup (T \cup W) = (S \cup T) \cup W$$

Fig. 4.4

For example, if S is the set of dogs and T is the set of horses, then $S \cup T$, the set of all objects which are either dogs or horses, is the same as $T \cup S$, the set of all objects which are either horses or dogs.

Proposition 3 corresponds to the commutative property of addition, that is, the sum of any numbers is the same, regardless of the order in which we add them. For any two numbers a and b, $a + b = b + a$.

Proposition 4. *If S, T, and W are any three sets, then*

$$(S \cup T) \cup W = S \cup (T \cup W).$$

Figure 4.4 shows a diagram representing S, T, and W.

Proof. The set $S \cup T$ consists of all objects which are elements of either S or T, hence $(S \cup T) \cup W$ consists of all those objects which are elements either of S or T, or of W. But $S \cup (T \cup W)$ consists of all those objects which are either in S or in $T \cup W$, hence $S \cup (T \cup W)$ consists of all those objects which are either in S, or in T or W. Therefore

$$(S \cup T) \cup W = S \cup (T \cup W).$$

For example, if S is the set of all dogs, T is the set of all horses, and W is the set of all fish, then $(S \cup T) \cup W$ is the set of all objects which either are dogs or horses, or are fish, while $S \cup (T \cup W)$ is the set of all objects which either are dogs, or are horses or fish.

Proposition 4 corresponds to the *associative* property of addition, that is, $(a + b) + c = a + (b + c)$ for any numbers a, b, and c.

Proposition 5. *If S, T, and W are any sets, then*

$$S \cap (T \cup W) = (S \cap T) \cup (S \cap W).$$

Proof. Recall that two sets A and B are equal if $A \subset B$ and $B \subset A$. We first show that $S \cap (T \cup W) \subset (S \cap T) \cup (S \cap W)$, that is, each element of $S \cap (T \cup W)$ is an element of $(S \cap T) \cup (S \cap W)$ as well. Let x be any element of $S \cap (T \cup W)$. Then x is an element of both S *and* $T \cup W$.

Since x is an element of $T \cup W$, either x is an element of T or x is an element of W. In any case, either x is an element of both S and T (if $x \in T$), or x is an element of both S and W. That is, x is an element of $S \cap T$, or x is an element of $S \cap W$. Therefore x is an element of $(S \cap T) \cup (S \cap W)$. Consequently, $S \cap (T \cup W) \subset (S \cap T) \cup (S \cap W)$.

Next we show that each element of $(S \cap T) \cup (S \cap W)$ is an element of $S \cap (T \cup W)$, that is, $(S \cap T) \cup (S \cap W) \subset S \cap (T \cup W)$. If x is any element of $(S \cap T) \cup (S \cap W)$, then x is either an element of $S \cap T$, or x is an element of $S \cap W$. If x is an element of $S \cap T$, then x is both an element of S and an element of T. On the other hand, if x is an element of $S \cap W$, then x is both an element of S and an element of W. In any case, x is an element of both S and either T or W. Therefore x is an element of $S \cap (T \cup W)$. Consequently, $(S \cap T) \cup (S \cap W) \subset S \cap (T \cup W)$. We therefore conclude that

$$S \cap (T \cup W) = (S \cap T) \cup (S \cap W).$$

We illustrate Proposition 5 in the next example.

Example 16. Let S be the set of all books in the living room, T be the set of books with green covers, and W be the set of novels. Then

$$S \cap (T \cup W)$$

is the set of all books in the living room which either have green covers or are novels.* Thus, $(S \cap T) \cup (S \cap W)$ expresses a set made up of two

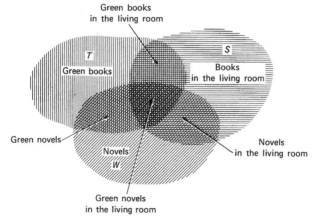

Green books
in the living room

T

Green books

S

Books
in the living room

Green novels

Novels
W

Novels
in the living room

Green novels
in the living room

Fig. 4.5

---

* We must always keep in mind that the "or" in such statements as these represents inclusive disjunction. Thus, an object in $S \cap (T \cup W)$ might be a novel with a green cover in the living room.

possibilities: $(S \cap T)$, the set of green covered books in the living room, or $(S \cap W)$ the set of novels in the living room. (See Fig. 4.5.) In this concrete situation, it is evident that $S \cap (T \cup W) = (S \cap T) \cup (S \cap W)$.

Note too that $S \cup (T \cap W)$ is the set of all objects which are either books in the living room or are both books with green covers and are novels; more concisely, $S \cup (T \cap W)$ is the set of books which are either in the living room, or are novels with green covers. On the other hand, $(S \cup T) \cap (S \cup W)$ is the set of books which either are in the living room or have green covers, *and* either are in the living room or are novels. The reader should analyze these statements to convince himself that in this instance we also have

10. $S \cup (T \cap W) = (S \cup T) \cap (S \cup W)$.

(See Exercises 1 through 4 which follow.)

Observe that Proposition 5 is similar to the distributive law of multiplication; that is, if a, b, and c are any numbers, then $a(b + c) = ab + ac$. Note that the statement corresponding to (10), that is,

$$a + (bc) = (a + b)(a + c)$$

is not true for numbers.

EXERCISES 4.4

1. Let $S = \{1, 2, 3, 4\}$, $T = \{1, 2, a, b\}$, and $W = \{3, a, d\}$. Verify each of the following.
   a) $S \cup T = T \cup S$ b) $S \subset S \cup T$
   c) $S \cap T \subset T$ d) $S \cap (T \cup W) = (S \cap T) \cup (S \cap W)$
   e) $S \cup (T \cap W) = (S \cup T) \cap (S \cup W)$ f) $S \cap T = T \cap S$
   g) $S \cap (T \cap W) = (S \cap T) \cap W$

2. Verify (a) through (g) in Exercise 1 with

   $$S = \{x \mid x \text{ is a student at Harvard}\},$$
   $$T = \{y \mid y \text{ is a student taking a history course at some college}\},$$
   and
   $$W = \{z \mid z \text{ is a student majoring in English at some college}\}.$$

3. Prove each of the following where S, T, and W are any sets.
   a) $S \cap T = T \cap S$ b) $S \subset S \cup T$
   c) $S \cap T \subset T$ d) $S \cap \varnothing = \varnothing$
   e) $S \cup \varnothing = S$ f) $S \cap (T \cap W) = (S \cap T) \cap W$
   g) $S \cup S = S$ h) $S \cap S = S$
   i) $S \cup (T \cap W) = (S \cup T) \cap (S \cup W)$

4. The following situation refers to Example 16. Analyze logically the following argument used to show that $(S \cup T) \cap (S \cup W) \subset S \cup (T \cap W)$. Suppose x is any book which is either in the living room or has a green cover, *and* which is also either in the living room or is a novel. If x is in the living room, then x is a book which is either in the living room or is a novel with a green cover. But if x is not in the living room, then x must both have a green cover and be a novel; therefore x must be a book which either is in the living room or is a novel with a green cover.

5. Prove each of the following (S, T, and W represent any sets).

   a) $S \subset T$ if and only if $S \cup T = T$. (See Example 15.)
   b) $S \subset T$ if and only if $S \cap T = S$.
   c) Suppose $S \subset T$. Then $S \cup W \subset T \cup W$.
   d) Suppose $S \subset T$. Then $S \cap W \subset T \cap W$.

6. If the intersection of two sets is compared with the multiplication of two numbers, what properties do intersection and multiplication share? What are the similarities between the relationship of $\varnothing$ to the intersection and union of sets and the relationship of zero to the addition and multiplication of numbers? A comparison was drawn between the properties of $\cup$ and the properties of $+$. Several similarities were found; now find several differences.

4.5 THE DIFFERENCE OF TWO SETS. COMPLEMENTS

As we have already seen, the intersection of the set S (all books in the living room) with the set T (all books with green covers) is the set of all green books in the living room. We might also wish to consider the set of all green books which are not in the living room, or the set of all books in the living room which do not have green covers. The latter set is called the *difference of S and T*; the former set is the difference of T and S.

Definition 4.5. *If S and T are any two sets, then the difference of S and T is defined to be the set of all objects which are elements of S, but not of T. We denote the difference of S and T by $S - T$. Thus,*

$$S - T = \{x \mid x \in S, \text{ but } x \notin T\}.$$

Similarly, $T - S$, the difference of T and S, is the set of all elements of T which are not elements of S; that is,

$$T - S = \{y \mid y \in T, \text{ but } y \notin S\}.$$

(See Fig. 4.6.)

In the special case that $T \subset S$, $S - T$ is called the *complement* of T in S. (See Fig. 4.7.)

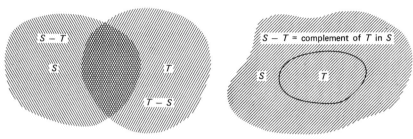

Fig. 4.6   **Fig. 4.7**

Example 17. Let $S = \{1, 2, 3, 4\}$ and $T = \{3, 4, a, b\}$. Then

$$S - T = (\mathrm{df})\ \{1, 2\}$$

and $T - S = (\mathrm{df})\ \{a, b\}$.

We should note that not only is $S - T$ unequal to $T - S$, but $S - T$ and $T - S$ share no elements whatever in common. For if x were an element of both $S - T$ and $T - S$, then x would have to be both an element of S which was not an element of T and an element of T which was not an element of S. Thus, x would have to be both an element of, and not an element of, T, a situation which is clearly impossible.

Example 18. Let S be the set of all cows, T be the set of all animals, and W be the set of all black objects. (See Fig. 4.8.) Then $S - T$ is the empty set since there are no cows which are not animals. On the other hand, $T - S$ is the set of all animals which are not cows, and is the complement of S in T since $S \subset T$.

We see that $S \cap W$ is the set of all black cows, while $S - W$ is the set of all cows which are not black. Since every cow is either black or not black, we have that $S = (S \cap W) \cup (S - W)$. We shall see that this equation is true in any similar application (Proposition 6).

We also note that $T - (S \cup W)$ is the set of all animals which are not elements of $S \cup W$ (that is, which are not either cows or black); thus,

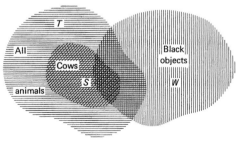

Fig. 4.8

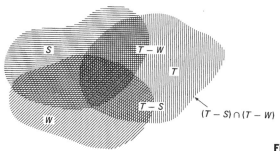

Fig. 4.9

$T - (S \cup W)$ is the set of animals which are neither cows nor black. On the other hand, $(T - S) \cap (T - W)$ is the set of objects which are both animals which are not black *and* animals which are not cows, a set that is exactly the same set as $T - (S \cup W)$. We shall see from Proposition 7 that this equality was no coincidence.

Proposition 6. *If S and T are any two sets, then* $S = (S \cap W) \cup (S - W)$.

Proof. We must show that

$$S \subset (S \cap W) \cup (S - W) \qquad \text{and} \qquad (S \cap W) \cup (S - W) \subset S.$$

If x is any element of S, then x is an element of S which is either in W (hence is in $S \cap W$), or is not in W (hence is in $S - W$). Therefore if $x \in S$, then $x \in S \cap W$, or $x \in S - W$, that is, $x \in (S \cap W) \cup (S - W)$. Consequently, $S \subset (S \cap W) \cup (S - W)$.

Now if x is an element of $(S \cap W) \cup (S - W)$, then either x is an element of $S \cap W$, or x is an element of $S - W$. In either case, x is an element of S. Therefore $(S \cap W) \cup (S - W) \subset S$. Consequently,

$$S = (S \cap W) \cup (S - W).$$

Proposition 7. *If S, T, and W are any sets (see Fig. 4.9), then*

a. $T - (S \cup W) = (T - S) \cap (T - W),$

and

b. $T - (S \cap W) = (T - S) \cup (T - W).$

Proof for (a). Suppose x is an element of $T - (S \cup W)$. Then $x \in T$, but $x \notin S \cup W$. Hence x is an element of T which is an element of neither S nor W. But then x is an element of both $T - S$ and $T - W$; that is, $x \in (T - S) \cap (T - W)$. We have therefore shown that

$$T - (S \cup W) \subset (T - S) \cap (T - W).$$

On the other hand, if x is an element of $(T - S) \cap (T - W)$, then x is an element of both $T - S$ and $T - W$. That is, x is an element of T but not of S, and x is an element of T but not of W. Therefore x is an element of T, but not an element of either S or W. In other words, $x \in T$ and $x \notin S \cup W$. Consequently, $x \in T - (S \cup W)$. We thus have that

$$(T - S) \cap (T - W) \subset T - (S \cup W).$$

Therefore

$$(T - S) \cap (T - W) = T - (S \cup W).$$

The proof of (b) is left as an exercise.

EXERCISES 4.5

1. Let $S = \{5, 6, 7, 8, 9\}$, $T = \{4, 5, 6, 11, A\}$ and $W = \{1, 5, A, B\}$. Verify each of the following.
   a) $S - T \subset S$
   b) $(S - T) \cap (T - S) = \varnothing$
   c) $T - (S \cup W) = (T - S) \cap (T - W)$
   d) $T - (S \cap W) = (T - S) \cup (T - W)$
   e) $(T - S) - W \neq T - (S - W)$
   f) $S = (S \cap W) \cup (S - W)$

2. Verify (a) through (f) in Exercise 1 when

$$S = \{x \mid x \text{ is a red house}\},$$
$$T = \{y \mid y \text{ is a building with twenty windows}\}$$

   and

$$W = \{w \mid w \text{ is a wooden object}\}.$$

3. Prove each of the following. S, T, and W are any sets.
   a) $S - S = \varnothing$
   b) $S - \varnothing = S$
   c) $S \subset T$ if and only if $S - T = \varnothing$
   d) $S = T$ if and only if $S - T = \varnothing$ and $T - S = \varnothing$
   e) $T - (S \cap W) = (T - S) \cup (T - W)$
   f) $\varnothing - S = \varnothing$
   g) $(S - T) \cap T = \varnothing$

4. Let S be the set of all people. What is the complement in S of each of the following sets?
   a) $\{x \mid x \text{ is a native of North America}\}$
   b) $\{y \mid y \text{ is a citizen of France}\}$
   c) $\{x \mid x \text{ is not a woman}\}$
   d) $\{y \mid y \text{ is not a citizen of Germany}\}$
   e) $\{w \mid w \text{ is neither a citizen of France, nor of Germany}\}$
   f) $\{z \mid z \text{ is both a professor and a citizen of France}\}$
   g) $\{x \mid x \text{ is a professor, but not a citizen of France}\}$

5. If S and T are sets and $T \subset S$, prove that $S - (S - T) = T$, that is, the complement of the complement in S of T is T.

6. Suppose S and T are sets. Is there necessarily a solution to each of the following equations?

a) $S \cup X = T$ b) $S \cap X = T$

c) $S - X = T$ d) $X - S = T$

Set Theory and Logic

5.1 SET DIAGRAMS

We used diagrams or figures to illustrate set-theoretic concepts on several occasions in Chapter 4. As a rule, the area inside a closed curve represents any given set. (See Fig. 5.1.) If it seemed more convenient, we could let the area of a triangle, rectangle, or other geometric configuration represent a set instead.

If we have several sets, each represented by a suitable area, we can indicate relationships among these sets by appropriate positioning of the areas representing the sets. For example, if we have sets S and T, then $S \subset T$ might be represented as in Fig. 5.2, while $S \cap T = \varnothing$ might be represented as in Fig. 5.3. Given two sets S and T without any assumptions as to how these sets are related to each other, we would generally position the areas representing S and T in the manner indicated in Fig. 5.4. Figure 5.4 also indicates the areas representing the various sets that can be formed from S and T.

Diagrams such as Fig. 5.4 are a great help in visualizing simple relationships between sets. If we proceed carefully, we may also draw conclusions from such diagrams.

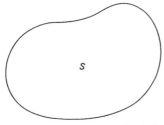

Fig. 5.1

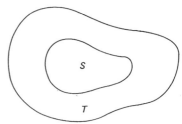

Fig. 5.2

Example 1. We note in Fig. 5.4 that the area representing $S - T$ added to the area representing $S \cap T$ gives us the total area representing the set S. But if we take the area which represents one set and add it to the area which represents another set, then the resulting area represents the union of the two sets. We thus conclude

1. $(S \cap T) \cup (S - T) = S.$

The relationship in (1) is an equality shown to be valid in Proposition 6 of Chapter 4.

   There are two things which should have disturbed the reader about Example 1. First, if the reader has had the stock experiences in his previous mathematics, he will have been warned about arguing from pictures. Pictures offer insights into a situation and may even lead the way to a proof, but a genuine proof must be based on established premises that lead to a desired conclusion by means of a logically valid argument. Second, the reader may wonder whether our diagram was based on the very conclusions we wished to prove. That is, if we constructed the diagram as we did because, even inadvertently, we assumed the truth of what we wished to prove, then obviously we shall be able to draw that conclusion from our diagram.

   The reader should review how (1) was proved in Proposition 6 of Chapter 4. We showed then that any element of $(S \cap T) \cup (S - T)$ was an element of S and that any element of S was an element of

$$(S \cap T) \cup (S - T).$$

We did this by applying the definitions of union, intersection, and the difference of two sets. To arrive at the portions of Fig. 5.4 which represent S, $S - T$, and $S \cap T$, we also apply these three definitions; moreover, the definitions are really all that we use. Thus, since $S - T$ is the set of elements of S which are not elements of T, that portion of Fig. 5.4 which represents $S - T$ should be all the area of S which is not common to the area representing T (that is, the area representing all elements of S which

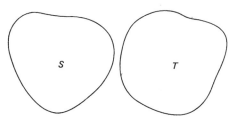

Fig. 5.3

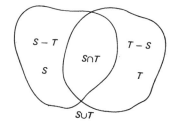

Fig. 5.4

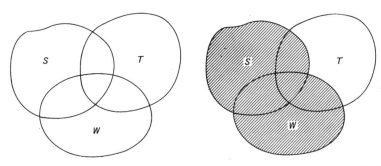

Fig. 5.5   **Fig. 5.6**

are not shared with T). The set $S \cap T$ is represented by the area common to that representing S and that representing T since this area represents all elements shared by S and T. Likewise, we see that $S \cup T$ is appropriately represented by adding the area representing S to the area representing T. Thus, our diagram (Fig. 5.4) is the result merely of applying the basic definitions of set theory and not of presupposing any propositions which must be proved.

However, there is still the question of whether we have a logically valid argument if we draw conclusions from a picture. For example, we could hardly say that we have proved that the sum of the interior angles of a plane triangle is equal to two right angles if we draw a picture of a triangle, measure its angles, and come up with 180 degrees; nevertheless, pictures do serve as a valuable aid in the study of plane geometry.

The fact is that in many instances, *if a diagram is properly used*, we can establish a conclusion to the satisfaction of almost everyone, even though technically we have not provided a formal proof. A diagram essentially remains an aid to finding a formal proof; but when it is sufficiently clear from a diagram how a formal proof should proceed, and if we could provide a formal proof on demand, then a diagram can serve in the place of the formal proof. The following example illustrates what we mean.

Example 2. Figure 5.4 should lead us at once to the following argument to prove (1): If x is an element of S, then either x is also in T (hence $x \in S \cap T$), or x is not in T (hence $x \in S - T$). Therefore if $x \in S$, $x \in (S \cap T) \cup (S - T)$. On the other hand, if $x \in (S \cap T) \cup (S - T)$, then either $x \in S \cap T$ or $x \in S - T$. In either case, $x \in S$. We have thus shown that $S \subset (S \cap T) \cup (S - T)$ and $(S \cap T) \cup (S - T) \subset S$; consequently, we have proved (1).

Practically speaking, however, if we wish to show that two sets are equal, it suffices for an informal (and usually convincing) proof to con-

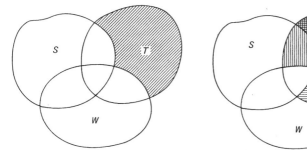

Fig. 5.7   **Fig. 5.8**

struct an appropriate diagram and then to show that the same area of the diagram represents both sets. The following example illustrates such a proof.

Example 3. We shall prove

2. $T - (S \cup W) = (T - S) \cap (T - W)$

if T, S, and W are any sets. We already proved this equality in Proposition 7 of Chapter 4. But now we can compare the formal proof in Chapter 4 with the informal diagrammatic proof presented here.

The appropriate diagram to start with is Fig. 5.5. In Fig. 5.6 we have shaded the area that represents $S \cup W$. The area that represents $T - (S \cup W)$ is the area of T which is totally outside the area representing $S \cup W$. This area of T is shaded in Fig. 5.7, and it represents the elements of T which are not elements of either S or W.

In Fig. 5.8 the area which represents $T - W$ is shaded with vertical lines, while the area which represents $T - S$ is shaded with horizontal lines. Thus the area which represents $(T - S) \cap (T - W)$ is the cross-hatched area (that is, the area common to the area representing $T - S$ and that representing $T - W$). This area is indicated in Fig. 5.9. We thus see that the area of Fig. 5.5 which represents $T - (S \cup W)$ is the same as the area which represents $(T - S) \cap (T - W)$. We therefore conclude the truth of (2).

Example 4 gives yet another example of this type of "proof by diagram."

Example 4. Let S and T be sets such that $S \subset T$. Then

3. $T - (T - S) = S$.

Here we have the hypothesis $S \subset T$, thus the correct diagram to begin

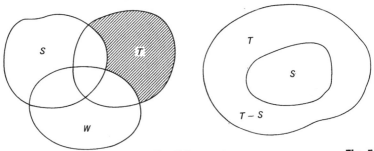

Fig. 5.9   **Fig. 5.10**

with is Fig. 5.2. In Fig. 5.10 we see that the area which represents $T - S$ is the area which represents all T that is not part of the area representing S. The area which represents $T - (T - S)$ is that area representing T which is not part of the area representing $T - S$, which is precisely the area representing S. Thus one portion of Fig. 5.10 represents both S and $T - (T - S)$, and hence we conclude that (3) is true.

EXERCISES 5.1

1. Label those portions of Fig. 5.5 which represent each of the following.
   a) S
   b) $S \cup T$
   c) $S \cap W$
   d) $W - S$
   e) $S \cap (T \cap W)$
   f) $S - (T - W)$
   g) $(S - T) - W$
   h) $(S \cup T) - W$

2. Suppose in Example 4 that we do not make the hypothesis that $S \subset T$. Then we should try to prove (3) by using Fig. 5.4. Show that without the hypothesis $S \subset T$ statement (3) need not be true.

3. Using the diagrams presented in Examples 3 and 4, construct formal proofs (2) and (3).

4. Prove each of the following by means of suitable diagrams. As usual, S, T and W will represent any sets.
   a) $S \cap T = T \cap S$
   b) $T \cap T = T$
   c) $S \cup (T \cup W) = (S \cup T) \cup W$
   d) $S \cap (T \cup W) = (S \cap T) \cup (S \cap W)$
   e) $T - (S \cap W) = (T - S) \cup (T - W)$
   f) If $S \subset T$, then $S \cup T = T$.
   g) If $S \subset T$ and $W \subset T$, then $S \cup T \subset W$
   h) If $S \subset T$, then $S - T = \varnothing$

5. Construct a suitable diagram for dealing with four arbitrary sets S, T, W, and Z. Using the diagram, prove each of the following.
   a) $S \cap (T \cap (W \cap Z)) = (S \cap T) \cap (W \cap Z)$
   b) $S - (T \cup (W \cup Z)) = (S - T) \cap ((S - W) \cap (S - Z))$

5.2 SETS AND LANGUAGE. VENN DIAGRAMS

Consider the statement

4. Bob plays both football and basketball.

We can express (4) in the language of sets as either

5. Bob is an element of the set of all objects which play both football and basketball

or

6. Bob is an element of both the set of objects which play football and the set of objects which play basketball.

We see then that Bob is an element common to $\{x \mid x$ plays football$\}$ and $\{y \mid y$ plays basketball$\}$, but this is the same as saying

7. Bob $\in \{x \mid x$ plays football$\} \cap \{y \mid y$ plays basketball$\}$.

Although (4) and (7) appear to be quite different, they say precisely the same thing, one using conversational language, the other, the language of sets. Note that "is" is translated by "is an element of" ($\in$) and "and" is translated by "$\cap$."

Actually, many statements can be translated into the language of set theory. The advantage of such a translation is that once we have expressed a statement in the language of sets, we can use what we know about sets to draw conclusions from the statement. For example, if we can express all the premises of an argument in the language of sets, we can use the theory of sets to determine whether or not the conclusion of the argument follows validly from the premises. (The reader should review Example 11 of Chapter 4.) The primary object of this section will be to discuss the translation of statements into the language of sets.

It might also have occurred to the reader that if we can express statements in the language of sets, then perhaps we can construct appropriate diagrams to aid us in drawing conclusions from these statements. Such is indeed the case. However, when the premises of an argument are under consideration, it is customary to assume that all the sets in question are subsets of an overall set, usually called the *universal set*. The universal set can be considered to be the set of all objects which we might be interested in during the course of the argument. We shall see that adding this universal set enables us to express and diagram certain statements more conveniently. In constructing a diagram to deal with statements, we shall usually let the universal set be represented by a rectangle within which we shall construct an appropriate set diagram. Letting U denote the

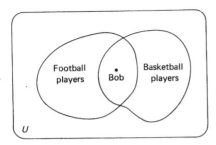

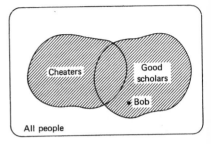

Fig. 5.11 **Fig. 5.12**

universal set, we would diagram (7) as shown in Fig. 5.11. Such a diagram is called a *Venn diagram*. We shall spend the remainder of this section discussing how to translate statements into the language of sets and how to construct an appropriate Venn diagram for each statement.

We have already seen that "and" is translated into set intersection. Now consider the statement

8. Bob is either a good scholar or cheats on his exams.

Statement (8) is equivalent to

9. Bob is either an element of the set of good scholars or an element of the set of all objects which cheat on their exams.

The universal set for a statement such as (9) could be the set of all people, that is, it might reasonably be assumed that in order for an object to be either a good scholar or a cheat that object would have to be a person. If some object is in one set or another, it is in the union of the two sets. Therefore in set language, (9) becomes

10. Bob $\in \{x \mid x$ is a good scholar$\} \cup \{y \mid y$ cheats on his exams$\}$.

The appropriate Venn diagram for (10) is given in Fig. 5.12. Note that although we know that Bob should be placed somewhere in the shaded portion of Fig. 5.12, we cannot say with certainty where he belongs.

Next consider the statement

11. Sam is not taking English.

Again, (11) might be considered in the context of the universal set of all people. Thus, (11) might then be rephrased either as "Sam is not an element of the set of people taking English," or as

12. Sam $\in U - \{x \mid x$ is taking English$\}$,

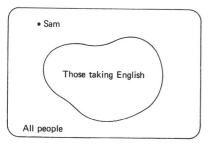

Fig. 5.13

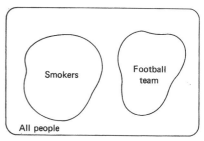

Fig. 5.14

where U is the set of all people. The appropriate Venn diagram is shown in Fig. 5.13.

Suppose now that we wish to translate the statement

13. No member of the football team smokes

into set language. Statement (13) is equivalent to

14. The set of members of the football team has no elements in common with the set of objects that smoke.

Once again the set of people can be considered to be the universal set. Two sets have no elements in common if and only if their intersection is the empty set $\varnothing$; thus (14) becomes

15. $\{x \mid x$ is a member of the football team$\} \cap \{y \mid y$ is a person who smokes$\} = \varnothing$.

An appropriate Venn diagram is given in Fig. 5.14.

Next, consider the statement

16. Some professors are good teachers.

Statement (16) can be rephrased as

17. The set of professors has at least one element in common with the set of good teachers.

Two sets have an element in common if and only if their intersection is not the empty set. Thus (16) becomes

18. $\{x \mid x$ is a professor$\} \cap \{y \mid y$ is a good teacher$\} \neq \varnothing$.

Once again the universal set might be the set of all people. An appropriate Venn diagram for (18) is shown in Fig. 5.15.

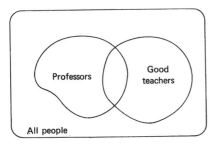

Fig. 5.15

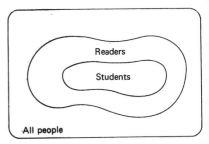

Fig. 5.16

The last type of statement we consider in this section is exemplified by

19. All students can read.

We have already seen that (19) can be expressed as

20. $\{x \mid x \text{ is a student}\} \subset \{y \mid y \text{ can read}\}$

(Example 11, Chapter 4). An appropriate Venn diagram for (20) is given in Fig. 5.16.

We should also note, however, that an "If ..., then ..." statement can often be expressed as an "All ... are ...," statement as we shall see in the following example. Conversely, an "All ... are ..." statement is really an "If ..., then ..." statement; for example, (19) is equivalent to "If someone is a student, he can read." This observation allows us to translate into the language of sets many statements which at first glance seem unsuited for translation.

Example 5. Consider the statement

21. If it is raining, I shall stay home.

Statement (21) is equivalent to

22. All of the time it is raining is time when I shall stay home.

Thus if we let the universal set be the set of all time, then (22), and hence (21), can be handled in the manner of (19).

Almost any statement can be expressed in the language of sets, but of course we must always be certain that the set translation expresses exactly what the original statement says, no more and no less. Likewise, in con-

structing a Venn diagram for any statement, we must be certain we do not
put anything into the diagram which is not implied by the original statement (or leave anything out that is implied, for that matter). There are
occasions when we cannot be sure exactly what Venn diagram is appropriate; for example, consider the discussion of (8). In such cases, we
must be particularly careful not to draw unjustifiable conclusions from
the diagram. We shall discuss this point at greater length in Section 5.3.

To conclude this section, we give below a table summarizing the ways
to translate certain ideas into set-theoretic language; in each case, we have
an example of a statement and its translation.

Table 5.1

Notion to be translated	Equivalent set notion	Type of Venn diagram	Sample statement	Translation of sample statement
Disjunction, "or," $\vee$	Union, $\cup$	Fig. 5.12	Mike will come or stay home.	Mike $\in \{x \mid x$ will come$\} \cup \{y \mid y$ will stay home$\}$
Conjunction, "and," $\wedge$	Inter-section, $\cap$	Fig. 5.11	Carol is smart and pretty.	Carol $\in \{x \mid x$ is smart$\} \cap \{y \mid y$ is pretty$\}$
Negation, $\sim$	Complement in universal set	Fig. 5.13	Mike is not handsome.	Mike $\in \{z \mid z$ is a person$\} - \{w \mid w$ is handsome$\}$
"No . . . is . . ."	Inter-section of two sets is $\varnothing$	Fig. 5.14	No student of English is stupid.	$\{x \mid x$ is a student of English$\} \cap \{y \mid y$ is stupid$\} = \varnothing$
"Some . . . are . . ."	Inter-section of two sets is not $\varnothing$	Fig. 5.15	Some students are wise.	$\{x \mid x$ is a stu-dent$\} \cap \{y \mid y$ is wise$\} \neq \varnothing$
"All . . . are . . ." "If . . . , then . . ."	Subset, $\subset$	Fig. 5.16	All men are mortal.	$\{w \mid w$ is a man$\} \subset \{y \mid y$ is mortal$\}$
"Is," "Are"	Dependent on the statement, no definite rule			

EXERCISES 5.2

1. Translate each of the following statements into the language of sets and draw an appropriate Venn diagram in each case. There may be several correct ways of translating some statements, as well as several possibilities for a choice of a universal set.

   a) Some men are soldiers.
   b) All tigers have tails.
   c) No one whom I know is going to the party.
   d) If Jack takes history, he will be smart.
   e) That animal is not a monkey.
   f) Math is both interesting and informative.
   g) English is either interesting or educational.
   h) Some men are not vegetarians.
   i) All men are wise and humane.
   j) Jack is neither interested nor creative.
   k) If it is snowing, then I shall wear my coat.
   l) Men are the only animals which do not have tails.
   m) All students are either smart or fail at least one course.
   n) The test will be hard only if either Mr. Jones or Mr. Smith makes it up.
   o) If it is raining, I shall not go outside.
   p) That Jack is taking English implies that he is either foolish or is a friend of the instructor.
   q) All men are not peaceful. (Is this statement ambiguous? Translate it according to both of its possible interpretations.)
   r) Sue is both creative and either interested or hardworking.
   s) Some criminals are not big men, but all criminals are dangerous.
   t) All tigers eat meat, and any wild animal that eats meat is dangerous.
   u) If it is either raining or cold, I shall not go to the party.
   v) Regardless of whether an integer is either even or odd, it is still positive, negative, or 0.
   w) All wild animals are dangerous, but some wild animals are kept in cages.
   x) No man is so strong that he cannot be tempted by money.

2. Below are pairs of statements; one statement in each pair is in conversational language and the other is in the language of sets. Decide whether the statements in each pair are equivalent. If they are not equivalent, determine if one of the statements implies the other.

   a) Either I saw Jack's new car, or I did not see it.

   Jack's car $\in \{x \mid x$ is a car I saw$\} \cup \{y \mid y$ is a car I did not see$\}$

   b) Steve is both a friend and a companion.

   Steve $\in \{x \mid x$ is a friend$\} \cap \{w \mid w$ is a companion$\}$

c) I am not the sort of person who would tell lies.

I $\notin \{w \mid w$ is a person who would tell lies$\}$

d) When Sam is not here, he is over at Sally's.

$\{x \mid x$ is a moment of time$\} \subset \{w \mid w$ is a moment of time when Sam is at Sally's$\} \cup \{z \mid z$ is a moment of time when Sam is not here$\}$

e) If Jack makes the track team, he will be happy; but if he does not, then he will drop out of school.

Jack $\in (\{x \mid x$ makes the track team$\} \cap \{y \mid y$ will be happy$\}) \cup (\{w \mid w$ does not make the track team$\} \cap \{z \mid z$ will drop out of school$\})$

f) If Sue is not here, she is home; but if she is here, she is not doing her homework.

Let U be the set of all people, W be the set of people at home, T be the set of people who are doing their homework, and V be the set of people who are here. Then

$$\text{Sue} \in ((U - V) \cap W) \cup (V \cap (U - T)).$$

3. It has been pointed out that in certain situations we cannot be sure exactly how the Venn diagram for a certain statement should be constructed. Draw all possible Venn diagrams for each of the following statements.

a) Jack is either a good scholar, or he cheats.
b) Some people have good manners.
c) Jack either has a lot of money, or he does not have a lot of money.

5.3 APPLYING SET THEORY TO ARGUMENTS

We have already discussed a number of argument forms in Chapter 3. In this section we shall consider some of these argument forms from a different point of view, and we shall also consider some new ones. Our technique will be to express the premises and conclusion of an argument in the language of sets and then use what we know about set theory to determine whether or not the conclusion follows necessarily from the premises. One example of this procedure was given in Example 11 of Chapter 4. We now give another example.

Example 6. A typical modus ponens argument is given by

23. If Alice marries Sam, she is crazy.

    Alice is marrying Sam.

    Alice is crazy.

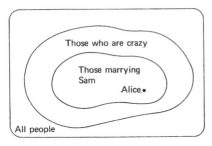

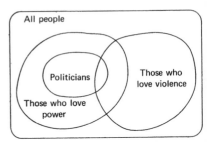

Fig. 5.17   **Fig. 5.18**

Expressed in the language of sets, (23) becomes

24. $\{x \mid x \text{ is Alice and } x \text{ marries Sam}\} \subset \{w \mid w \text{ is crazy}\}$;
that is, $(\{\text{Alice}\} \cap \{x \mid x \text{ marries Sam}\}) \subset \{w \mid w \text{ is crazy}\}$

Alice $\in \{x \mid x \text{ marries Sam}\}$; that is, $\{\text{Alice}\} \subset \{x \mid x \text{ marries Sam}\}$

Alice $\in \{w \mid w \text{ is crazy}\}$; that is, $\{\text{Alice}\} \subset \{w \mid w \text{ is crazy}\}$

An appropriate Venn diagram based on the premises of (24) is shown in Fig. 5.17. From set-theoretic considerations, we see that the conclusion of (24) does indeed follow from the premises. From the second premise we obtain

$$\{\text{Alice}\} \cap \{x \mid x \text{ marries Sam}\} = \{\text{Alice}\} ;$$

thus the first premise gives

$$\{\text{Alice}\} \cap \{x \mid x \text{ marries Sam}\} = \{\text{Alice}\} \subset \{w \mid w \text{ is crazy}\},$$

which is the conclusion. The argument is therefore valid.

We have already had some experience in dealing with arguments in which one or more of the premises was of the form "All . . . are . . . ," for example, "All men are mortal." Thus far, however, we have done little with arguments in which some premise is of the form "Some . . . are . . . ," for example, "Some men are wise." Now consider the following argument:

25. All politicians love power.

Some men who love power love violence.

Some politicians love violence.

From a set-theoretic point of view, (25) can be expressed as

26. $\{x \mid x \text{ is a politician}\} \subset \{w \mid w \text{ loves power}\}$

$\{w \mid w \text{ loves power}\} \cap \{z \mid z \text{ loves violence}\} \neq \varnothing$

$\{x \mid x \text{ is a politician}\} \cap \{z \mid z \text{ loves violence}\} \neq \varnothing$

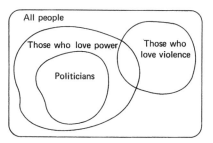

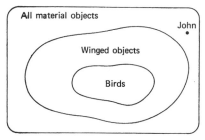

Fig. 5.19   Fig. 5.20

If we now try to construct an appropriate Venn diagram for (26), we see that the premises of (26) do not help us decide which of Figs. 5.18 or 5.19 expresses the real state of affairs. Yet both figures concur with the premises of (26). From Fig. 5.19 we see that the conclusion of (26) does not follow necessarily from the premises of (26), hence (26) is an invalid argument.

We conclude this section by analyzing a modus tollens argument by means of set theory.

Example 7

27. All birds have wings.

John does not have wings.

John is not a bird.

First we express (27) in the language of sets. One way of doing this is as follows:

28. $\{x \mid x \text{ is a bird}\} \subset \{y \mid y \text{ has wings}\}$

John $\notin \{y \mid y \text{ has wings}\}$

John $\notin \{x \mid x \text{ is a bird}\}$

Figure 5.20 gives an appropriate Venn diagram for (28). We thus conclude that (27) is a valid argument.

EXERCISES 5.3

1. First, express each of the following arguments in the language of set theory. Then construct an appropriate Venn diagram to illustrate the situation described by the premises of each argument. If more than one Venn diagram is consistent with the premises of an argument, as was true for (26), construct as many Venn diagrams as are consistent. Determine in each instance whether the argument is valid or invalid.

a) All men are ambitious.

Anyone who is ambitious is doomed to disappointment.

All men are doomed to disappointment.

b) All tigers are dangerous.

Some things which are dangerous are found in New York City.

Some tigers are found in New York City.

c) There is no man who has not wanted to be king.

There is no king who is happy.

There is no man who has wanted to be happy.

d) No tigers are found in Kansas.

Some steers are found in Kansas.

Some steers are not tigers.

e) No gold is found in Texas.

All oil is found in Texas.

No gold is oil.

f) If Sam robs the bank, he will be caught.

If anyone is caught, he will go to jail.

If Sam robs the bank, he will go to jail.

2. Analyze each of the following arguments using set theory. Begin by breaking the argument down into elementary arguments.

a) If Sam goes big game hunting, he will get hurt. If he gets hurt, he will be sorry that he went. But if he is sorry that he went, he will be mad at you for suggesting it. Therefore if Sam goes big game hunting, he will be mad at you.

b) There are no tigers in Africa. Today I saw a tiger. Therefore today I am not in Africa. But if I am not in Africa, I am either in New York or New Jersey. If I were in New Jersey, I would be unhappy. I am not unhappy. Therefore I am in New York.

c) I do not believe some of the things I hear. But if Sam tells me something, then I believe it. Sam did not tell me that you will be here today. Therefore I do not believe that you will be here today.

chapter six

Counting

6.1 THE NOTION OF COUNTING

Suppose we are given a box containing apples and oranges. How can we tell whether the number of apples in the box is the same as the number of oranges? Offhand this does not seem like a particularly taxing problem. The obvious answer is that we count the number of apples and then count the number of oranges, and if in each case we obtain the same number, we know there are as many oranges as apples.

But now suppose we did not know how to count, or that we did not know what a number is. If the reader feels that these suppositions are absurd, let him examine his own notions of number and counting to see whether they are merely vague, intuitive, or mechanical notions, or whether they are clearly defined ideas. The truth is, of course, that we can and do use numbers without knowing precisely what they are. Most bookkeepers would have little interest in a theoretical discussion of numbers and counting, while on the other hand, many theoretical mathematicians have trouble balancing their checkbooks.

Nevertheless, the concept of number is one of the most fundamental concepts in all of mathematics. There are few areas of mathematical study into which the number concept does not enter. Strangely enough, however, it is only comparatively recently, in fact in this century, that a satisfactory notion of number was found. Of course, we assume that the reader can add, subtract, multiply, and divide; he has been schooled in the mechanics of these operations since early childhood. In this chapter we shall not concern ourselves with mechanical procedures, but rather with the underlying question: What is a number?

Before returning to the box of apples and oranges, let us consider another situation. Suppose we have a crowd of people and a certain number of chairs. What would be a simple test to determine whether we have the same number of chairs as people? We can have as many people

Fig. 6.1

as possible sit down in the chairs, only one person to a chair. If everyone finds a chair and there are no vacant chairs, then the number of people is exactly the same as the number of chairs. If everyone has a chair, but there are some vacant chairs, then there are more chairs than people. If after all the chairs are filled, some people are still standing, then there are more people than chairs.

What we did, essentially, was to pair one person with precisely one chair, and continue the pairing process until the chairs or the people ran out. If both chairs and people ran out at the same time, then we have the same number of chairs as people.

What we might do with the oranges and apples is to pair one apple with precisely one orange and continue the pairing process until either all the apples or all the oranges were used up. (Fig. 6.1.) If the oranges and apples were both used up at the same time, then we would have the same number of oranges as apples.

Perhaps at this point the reader can't help wondering whether the author has lost his mind. Anyone with a modicum of intelligence is able to determine whether two sets have the same number of elements, or at least he would be able to do so if he had the time to actually count the number of elements in each set—something that might take a while if the sets were very large. But certainly if there were not sufficient time to count the sets, there would probably not be time enough to pair the elements of the sets either. Three observations should help put this study in its proper perspective.

1.   When we count any set, we are really pairing the elements of the set with positive integers. For example, suppose we have a bag of oranges and we proceed to count the oranges. We pick some orange and say, "One." We set the first orange aside, pick another orange and say, "Two." We have paired an orange with the integer 1, another orange with the integer 2. As we proceed with the counting, we pair each orange with precisely one positive integer, and of course we take the integers in their usual order. If we end up pairing all the oranges with the elements of the set $\{1, 2, 3, \ldots, 10\}$, that is, if the last orange counted is paired with 10, then we say we have 10 oranges. If there are 109 oranges, then we shall have paired the oranges with the elements of the set $\{1, 2, \ldots, 108, 109\}$.

Now if we wish to see whether two sets S and T contain exactly the same number of elements, we can count both sets, pairing each set with a set of integers in a special way; that is, we pair each element of S with precisely one element of a set of consecutive positive integers, say $\{1, 2, \ldots, n\}$, thus each element of $\{1, 2, \ldots, n\}$ has been paired with some one element of S. We then say that S has n elements. Likewise, we repeat the procedure for T, where the appropriate set of positive integers is $\{1, 2, \ldots, m\}$. Then if $n = m$, that is, if

$$\{1, 2, \ldots, n\} = \{1, 2, \ldots, m\},$$

we say that S and T have the same number of elements.

Our first point, therefore, is that, mathematically, counting two sets is actually more complicated than merely pairing the elements of the two sets directly. Just as it is cheaper to buy goods from a wholesaler and eliminate the middleman, so mathematically it is "cheaper" to pair two sets directly than to pair each set with some subset of the positive integers. Even though our schooling and certain considerations of convenience make counting more practical, it is in fact a more efficient procedure to pair two sets directly.

2. We want to work toward a definition of number which is not dependent on our already knowing what a number is. We will soon define the notion of "having the same number of elements" without any appeal to a prior knowledge of numbers. This in turn will help us to discuss numbers without the risk of circular reasoning.

3. While we may be able to count finite sets quite conveniently, we shall find it impractical to count infinite sets, that is, sets which are not finite. Our definition of "has the same number of elements as" will have meaning, however, even when we are dealing with infinite sets. We shall find, in fact, that the study of infinite sets leads to some interesting (and possibly startling) results.

Following the lead of the examples given earlier in this section, we make the following definition.

Definition 6.1. *Two sets S and T are said to have the same number of elements if each element of S can be paired with precisely one element of T in such a way that every element of T is paired with precisely one element of S.*

Example 1. The sets $\{A, B\}$ and $\{1, 2\}$ have the same number of elements. In particular, we may set up the following pairing: $A \leftrightarrow 1$, $B \leftrightarrow 2$, where the symbol $\leftrightarrow$ will denote the phrase "is paired with." Note that each element of $\{A, B\}$ has been paired with precisely one element of $\{1, 2\}$ and each element of $\{1, 2\}$ is paired with precisely one element of $\{A, B\}$.

This pairing establishes that $\{A, B\}$ and $\{1, 2\}$ have the same number of elements. Another pairing which would have worked just as well is $A \leftrightarrow 2$ and $B \leftrightarrow 1$.

Note that the pairing $A \leftrightarrow 1$ and $B \leftrightarrow 1$ would not be sufficient to show that $\{A, B\}$ and $\{1, 2\}$ have the same number of elements. For although each element of $\{A, B\}$ is paired with precisely one element of $\{1, 2\}$, the element 2 of $\{1, 2\}$ has not been paired with any element of $\{A, B\}$; moreover, the element 1 of $\{1, 2\}$ has been paired with two elements of $\{A, B\}$ and not just one.

Example 2. The sets $\{E, F, G\}$ and $\{6, 7\}$ do not contain the same number of elements. For no matter how we might try to pair the elements of $\{E, F, G\}$ with the elements of $\{6, 7\}$, we shall always be forced to pair two elements of $\{E, F, G\}$ with some one element of $\{6, 7\}$. That is, we cannot find a pairing of the sort needed in order to establish that $\{E, F, G\}$ and $\{6, 7\}$ have the same number of elements.

In Examples 1 and 2 (as well as in the previous discussion), the sets under consideration were finite. Definition 6.1, however, applies equally well to infinite sets, as we shall see in the following example.

Example 3. Consider the set $N = \{1, 2, 3, 4, \ldots\}$ of all positive integers and the set $M = \{2, 4, 6, 8, \ldots\}$ of all positive even integers, that is, the set of all positive multiples of 2. We shall show that N and M have the same number of elements. We note that M is a subset of N, hence we have the somewhat surprising result that when we are dealing with an infinite set, we can discard some elements and still wind up with as many elements as we had to start with. We can pair the elements of N with the elements of M as indicated below:

$$
\begin{array}{ccccccccc}
N: & 1 & 2 & 3 & 4 & 5 & 6 & \ldots & n & \ldots \\
& \updownarrow & \updownarrow & \updownarrow & \updownarrow & \updownarrow & \updownarrow & & \updownarrow & \\
M: & 2 & 4 & 6 & 8 & 10 & 12 & \ldots & 2n & \ldots
\end{array}
$$

That is, the positive integer n is paired with the positive even integer $2n$. It should be clear that each positive integer is paired in this way with exactly one positive even integer. On the other hand, if k is any positive even integer, then k is a multiple of 2; that is, $k = 2n$ for some positive integer n. Then according to our pairing, k has been paired with the positive integer n and only the positive integer n; that is, each positive even integer has been paired with precisely one positive integer. We are therefore entitled to say that M and N have the same number of elements.

The reader should carefully note that the previous example included several pairings but not all the individual pairings explicitly. Instead, we learned a rule so that, presented with any positive integer, we could immediately establish what positive even integer it should be paired with. If we tried to "count" the set M of Example 3, we would find it an impossible task since we could never exhaust all the elements of M. Nevertheless, we can establish a pairing between the sets M and N by finding a rule or pattern which clearly describes in each instance what the pairing should be.

It often happens that students who are studying infinite sets for the first time are confused about the properties of such sets. Their difficulties usually stem from the haziness of their prior notions of the number concept. We shall soon see that two infinite sets may indeed not have the same number of elements; that is, there are different kinds of infinity.

EXERCISES 6.1

1. Prove that the first set in each of the following pairs of sets has the same number of elements as the second set. This, of course, must be done by finding a pairing of the elements of the first set with the elements of the second set in such a way as to satisfy Definition 6.1. There may be many correct pairings.

   a) $\{G, 7\}$ and $\{K, 6\}$
   b) $\{4, 5, 6, 7\}$ and $\{0, 1, 2, 3\}$
   c) $\{z \mid z$ is a letter of the English alphabet$\}$ and $\{1, 2, 3, \ldots, 26\}$
   d) the set N of positive integers and the set K of positive multiples of 3 [*Hint:* Use the type of pairing that was used in Example 3.]
   e) $\{z \mid z$ is an integer greater than or equal to 0$\}$ and the set N of positive integers. [*Hint:* Write a few elements of N in a line as shown below and underneath write the elements of the first set in their natural order, that is,

   $$\begin{array}{ccccc} 1 & 2 & 3 & 4 & 5 \; \ldots \\ 0 & 1 & 2 & 3 & 4 \; \ldots \; ; \end{array}$$

   try pairing 0 with 1, 1 with 2, 2 with 3, etc. What is the general rule then for pairing elements of the first set with elements of the second?]

2. Prove that any set S has the same number of elements as itself. [*Hint:* Pair each element of S with itself.]

3. Recall that $\varnothing$ is the set which contains no elements whatever. Prove that if S is a nonempty set (that is, S contains at least one element), then S and $\varnothing$ do not have the same number of elements.

4. Below are certain statements which appear to be similar to Definition 6.1. Actually, each does not say the same thing as Definition 6.1; that is, none of the following could be used to define "has the same number of elements." Prove this in each instance by finding an example of two sets, together with a pairing of the elements of the two sets, which satisfies the given statement, but does not satisfy Definition 6.1.

a) Two sets S and T have the same number of elements if each element of S can be paired with precisely one element of T in such a way that each element of T is paired with some element of S. [*Hint:* Try $S = \{1, 2\}$, $T = \{A\}$, and the pairing $1 \leftrightarrow A$ and $2 \leftrightarrow A$.]

b) Two sets S and T have the same number of elements if each element of S can be paired with some element of T in such a way that each element of T is paired with precisely one element of S.

c) Two sets S and T have the same number of elements if each element of S can be paired with precisely one element of T.

6.2 MORE ABOUT INFINITY

There is probably no concept in mathematics which has caused more consternation among mathematicians, both novices and experts, and among philosophers too for that matter, than the concept of infinity. From the time of the ancient Greeks until the present, some of the most searching and fundamental study in mathematics has been concerned with infinity. It is beyond the scope of this text to delve very deeply into even some of the simpler problems relating to infinite sets. Nevertheless, we shall devote the next two sections to developing some of the observations about infinity that were made in Section 6.1.

First we must note that in Section 6.1 we defined neither the notion of an infinite set nor the notion of a finite set. We intuitively think of a finite set as a set which has the same number of elements as a set of the form $\{1, 2, 3, \ldots, n\}$ for some appropriate positive integer n. However, this concept of a finite set rests in turn on our concept of the positive integers. These assumptions imply that our notion of finiteness rests on our notion of number, and it is this notion of number that we are trying to formulate and develop in this chapter.*

In other words, it appears that our notion of a finite set depends on some prior notion of what a number is, or at least what an integer is. The fact is, however, that we can develop the theory of the integers without appealing to the notion of number. Thus, the use of the notion of finiteness to investigate the notion of number is not really circular. Even so, since we have not developed the concept of integers rigorously but are still relying on an intuitive notion of the integers, our notion of finiteness remains essentially intuitive; consequently, our notion of infinity remains intuitive as well.

Instead of defining "finite" and then defining "infinite," we could try to define "infinite" and then say that "finite" was not-infinite. Indeed, this can be done. Specifically, we can define a set S to be infinite if S con-

---

* Note that we have not yet defined what we mean by *number*, only what we mean by *has the same number of elements as*.

tains a subset T, $S \neq T$, such that S and T have the same number of elements. This approach, while satisfactory from the point of view of rigor, involves us with techniques beyond the ambitions of this book.

Our solution to this dilemma will be to presume that the notion *finite*, like the notion of integer, is intuitively understood (at least well enough so that the reader can recognize a finite set when he sees one).

We now show that two infinite sets may not have the same number of elements.

Example 4. Let N be the set of positive integers and T be the set of all unending decimals whose digits are either 0 or 1. For example, a typical element of T would be .00111000 ... (The dots indicate that the decimal does not terminate.) We now show that N and T do not have the same number of elements. If N and T did have the same number of elements, then there would have to be a pairing, (which we will denote by π) of the elements of N with the elements of T in accordance with Definition 6.1.

Table 6.1

n	Element of T paired with n by π
1	.001101 ...
2	.111011 ...
3	.110111 ...
4	.111111 ...
...	...
...	...

That is, the pairing π would be such that each element of N is paired with precisely one element of T and each element of T is paired with precisely one element of N. If there is in fact such a pairing, then we can construct a table such that a positive integer is given in one column and the decimal it is paired with is placed opposite it in another column (Table 6.1). We now show that this pairing which is supposed to satisfy Definition 6.1 cannot satisfy that definition. In particular, we shall construct an element of T which has not been paired with any element of N. We construct the decimal

$$D = .d_1 d_2 d_3 d_4 \ldots$$

(where d_1 stands for the first digit of D, d_2 for the second digit, etc.) as follows: If the decimal paired with 1 has 1 as its first digit, we make d_1 be 0; if the decimal paired with 1 has 0 as its first digit, we set $d_1 = 1$. Then D differs from the decimal paired with 1 at least in the first digit. If the decimal paired with 2 has 1 as its second digit, set $d_2 = 0$; if the

paired with 2 has 0 as its second digit, set $d_2 = 1$. Then D differs from
the decimal paired with 2 at least in the second digit. As a rule, if the
decimal paired with the integer n has 1 as its nth digit, set $d_n = 0$; if the
decimal paired with n has 0 as its nth digit, set $d_n = 1$.

Since D is an unending decimal that uses only either 0 or 1 for its
digits, D is an element of T. And since we supposed that we had a pairing
which proved that N and T had the same number of elements, D must
have been paired with some positive integer. But D was not paired with
1 since it differs from the decimal paired with 1 at least in the first digit.
Similarly, D was not paired with 2. In fact, D cannot be paired with any
integer n since by construction D differs from the decimal paired with n
at least in the nth digit. We have thus produced an element of T which is
not paired with any element of N. This contradicts our assumption that
we had a pairing which showed N and T to have the same number of
elements. The assumption that there is a pairing which shows N and T
to have the same number of elements is therefore false; consequently, N
and T do not have the same number of elements.

The following definition proves useful in discussing infinite sets.

Definition 6.2. *A set S is said to be countable if S contains the same number
of elements as some subset of the set of positive integers.*

Any finite set is countable. Since the set N of positive integers is a
subset of itself (Chapter 4, Proposition 1) and N has the same number
of elements as itself (Section 6.1, Exercise 2), the set N itself is countable.
We also see from Example 3 that the set M of even positive integers is
countable. We shall see in the next section that the set T of Example 4
is not a countable set.

Before developing further the notion of a countable set, we shall prove
two fundamental properties of "has the same number of elements as."
To do so, we shall use the following notation. If π is a pairing of the
elements of a set S with the elements of a set T and the element s of S is
paired in π to the element t of T, we shall write $s \overset{\pi}{\leftrightarrow} t$.

Proposition 1. *If a set S has the same number of elements as a set T, then
the set T has the same number of elements as S.*

Proof. Since S has the same number of elements as T, we can select a
pairing between the elements of S and the elements of T in accordance
with Definition 6.1. We define a pairing between the elements of T and
the elements of S as follows: If s is paired with t by the selected pairing π,
then pair t with s. That is, if $s \overset{\pi}{\leftrightarrow} t$ (in the pairing of the elements of S
with the elements of T), then t is paired with s to form the pairing of the

elements of T with the elements of S.* If the original pairing π satisfied the conditions of Definition 6.1, then the new pairing will also satisfy the conditions. Specifically, since π had each element of T paired with a unique element of S, the second pairing also has this property. And since the first pairing had every element of S paired with a unique element of T, the second pairing also has this property. Therefore, T has the same number of elements as S.

Proposition 2. *If a set S has the same number of elements as a set T, and T has the same number of elements as the set W, then S has the same number of elements as W.*

Proof. Since S has the same number of elements as T, we can select a pairing π of the elements of S with the elements of T in accordance with Definition 6.1; suppose the element s of S is paired in π with the element t of T. Since T has the same number of elements as W, we can select a pairing π' of the elements of T with the elements of W in accordance with Definition 6.1; suppose t is paired with w in π'. Then pair s with w. That is, if

$$s \overset{\pi}{\leftrightarrow} t \qquad \text{and} \qquad t \overset{\pi'}{\leftrightarrow} w,$$

then pair s with w to obtain a pairing of the elements of S with the elements of W. The pairing between the elements of S and the elements of W defined in this manner can be shown to satisfy the conditions of Definition 6.1; proof of this is left as an exercise.

EXERCISES 6.2

1. Show that the pairing between the elements of S and the elements of W defined in the proof of Proposition 2 satisfies the conditions of Definition 6.1.

2. In order to understand better the procedure used in Propositions 1 and 2, find a pairing which shows that the first set in each of the following triples of sets has the same number of elements as the second set in the triple; then find a pairing which shows that the second set in the triple has the same number of elements as the third set in the triple. Next construct the pairing (based on the particular pairing you constructed between the first and second sets) which shows that the second set has the same number of elements as the first set (according to the procedure described in the proof of Proposition 1). Then find the pairing between the elements of the first set and the elements of the third set (in accordance with the procedure of the proof of Proposition 2) which shows that these sets have the same number of elements. *Illustration:* $\{3, 4\}$, $\{5, 6\}$, $\{8, 9\}$. A pairing of the first set with the second

---

* Cf. Exercise 2 of this section.

set is $3 \leftrightarrow 5$ and $4 \leftrightarrow 6$; a pairing of the second set with the third set is $5 \leftrightarrow 8$ and $6 \leftrightarrow 9$. The pairing of the second set with the first set then is $5 \leftrightarrow 3$ and $6 \leftrightarrow 4$, and the pairing of the first set with the third set is $3 \leftrightarrow 8$ and $4 \leftrightarrow 9$.

a) $\{A, B, C\}$, $\{a, b, c\}$, $\{1, 2, 3\}$
b) $\{3, 4, 5, 6\}$, $\{5, 6, 7, 8\}$, $\{7, 8, 9, 10\}$
c) $\{A, B, C, D, E\}$, $\{A, B, C, D, F\}$, $\{A, B, C, F, G\}$
d) The set of positive integers, the set of positive even integers, the set of positive multiples of 3
e) The set of integers greater than or equal to 0, the set of integers greater than or equal to -1, the set of integers greater than or equal to -5

3. Prove that the set T of Example 4 has the same number of elements as the set of all subsets of the positive integers. [*Hint:* Let W be a subset of N, the set of positive integers. Pair W with the decimal $.d_1 d_2 d_3 d_4 \ldots$, where $d_1 = 1$ if 1 is in W, and $d_1 = 0$ if 1 is not in W; $d_2 = 1$ if 2 is in W, and $d_2 = 0$ if 2 is not in W; etc. Show that this pairing satisfies the conditions of Definition 6.1.] We know that the set of positive integers is countable, but we will soon show that T is uncountable. We therefore conclude that a countable set may have uncountably many subsets.

4. Suppose S and T are any sets. Suppose S contains n elements and T contains m elements. What is the smallest number of elements that $S \cup T$ can contain? What is the largest number of elements that $S \cup T$ can contain? What are the smallest and largest numbers of elements that $S \cap T$ can contain? Find a formula for the number of elements of $S \cup T$ in terms of the number of elements of S, T, and $S \cap T$.

5. Prove that the union of two finite sets is finite. Prove that any subset of a finite set is a finite set.

6.3 MORE ABOUT COUNTABLE SETS

Proposition 3. *If W is any infinite subset of the set N of positive integers, then W has the same number of elements as N.*

Proof. Arrange the elements of W in their natural order. Pair the first element of W with 1, the second element of W with 2, the third element of W with 3. In general, pair the nth element of W with the positive integer n. We now show that this pairing satisfies the conditions of Definition 6.1. First, each element of W is paired with precisely one positive integer, that is, precisely one element of N. Moreover, if n is any positive integer, then since W is infinite, W contains an nth element (which is unique). Therefore each element of N is paired with precisely one element of W. Hence W and N have the same number of elements.

Corollary 1. *A set T is countable if and only if T has the same number of elements as the set of positive integers or is finite (we consider the empty set to be a finite set).*

Proof. If T is a countable set, then T has the same number of elements as a subset W of N. If $W = \{1, 2, 3, \ldots, n\}$ for some positive integer n, then T is finite. However, suppose W is an infinite subset of N. Then by Proposition 3, W has the same number of elements as N. Since T has the same number of elements as W, and W has the same number of elements as N, then by Proposition 2, T has the same number of elements as N.

Corollary 2. *The set T described in Example 4 is not countable.*

Proof. If the set T of Example 4 were countable, then T would have to be finite (which it obviously is not), or by Corollary 1, T would have to have the same number of elements as N, which was shown in Example 4 to be impossible.

Example 5. The set T of decimals described in Example 4 forms only part of all the numbers between 0 and 1, inclusive. We might therefore suspect that

$$S = \{x \mid x \text{ is a number between 0 and 1, inclusive}\}$$

is also an uncountable set. Such is indeed the case. Let us suppose that S is countable: since S is not finite, S must contain the same number of elements as N, the set of positive integers. This means, however, that there is a pairing π of the elements of S with the elements of N in such a way that each element of S is paired in π with exactly one element of N and each element of N is paired with exactly one element of S. But T is a subset of S. Let W be the positive integers that have been paired in π with elements of T. Then each element of T is paired with precisely one element of W and each element of W is paired with precisely one element of T; that is, T has the same number of elements as W, a subset of N. Consequently, T is countable. But this contradicts the fact that T was shown to be uncountable. The supposition that S is countable is therefore false; hence S is uncountable.

An argument almost identical to the above can be used to show that the full set of real numbers is uncountable. Let us look at the next proposition.

Proposition 4. *If a set S contains an uncountable subset T, then S itself is uncountable.*

We leave the proof as an exercise for the reader. This proof should be modeled after the argument of Example 5. During the course of Example 5, we showed essentially that if S is a countable set, then any subset of S is countable. We formalize this finding in the next proposition.

Proposition 5. *If S is any countable set and T is a subset of S, then T is also a countable set.*

Proof. If S is finite, then any subset of S is also finite (Section 6.2, Exercise 5), and hence is countable. Suppose S is infinite. Since S is countable we can select a pairing π of the elements of S with the positive integers in accordance with Definition 6.1. Let W be the set of positive integers which are paired in π with elements of T. Then T and W have the same number of elements. Consequently, since W is a subset of the positive integers, T is countable.

If S and T are any two countable infinite sets, then both S and T have the same number of elements as N, the set of positive integers; that is, S has the same number of elements as N, and T has the same number of elements as N. By Proposition 1, then, N has the same number of elements as T. Therefore since S has the same number of elements as N and N has the same number of elements as T, S has the same number of elements as T by Proposition 2. We are therefore entitled to say that any two countable infinite sets have the same number of elements. We might ask whether any two uncountable sets have the same number of elements. The answer to this question is no, but to prove it would take us beyond the scope of this text. There are, in fact, not just two or three kinds of infinite sets, but there are infinitely many kinds of infinity. More precisely, we can find an infinite collection of infinite sets, no two of which have the same number of elements!

Any subset W of the set N of positive integers is countable (since W has the same number of elements as W itself, which is a subset of N). If U and V are any two subsets of the positive integers, then $U \cup V$ is also a subset of the positive integers, and hence is countable. That is, the union of any two subsets of the positive integers is a countable set. The following proposition shows that the union of any two countable sets (whether or not they be subsets of the positive integers) is again a countable set.

Proposition 6. *If S and T are countable sets, then $S \cup T$ is a countable set.*

Proof. If S and T are both finite, then $S \cup T$ is finite, and hence is countable (Section 6.2, Exercise 5). We next consider the following cases.

CASE 1. S is finite and T is infinite. Since S is finite, S has the same number of elements as $\{1, 2, \ldots, n\}$ for some positive integer n. Now consider $T - S$, the set of all elements of T which are not elements of S. Since T is infinite and S is finite, $T - S$ must be an infinite set (for if $T - S$ were finite, then T would be a subset of the finite set $S \cup (T - S)$ and hence would be finite, a contradiction). Then, since $T - S$ is a subset of the countable set T, $T - S$ is itself countable (Proposition 5); consequently, $T - S$ has the same number of elements as any infinite

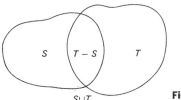

Fig. 6.2

subset of the set N of positive integers. In particular, $T - S$ has the same number of elements as the set of all integers larger than n, that is,

$$\{n + 1,\, n + 2,\, n + 3, \ldots\}.$$

We know that S and $T - S$ share no elements in common; moreover, $S \cup (T - S) = S \cup T$ (see Fig. 6.2). In accordance with Definition 6.1, there is a pairing of S with $\{1, 2, \ldots, n\}$ and there is a pairing of $T - S$ with $\{n + 1, n + 2, \ldots\}$. "Combining" these two pairings gives us a pairing between $S \cup T$ and N which shows that $S \cup T$ and N have the same number of elements. The manner of combining the pairings can best be outlined by the arrangement below. The two pairings together pair $S \cup T$ with N in accordance with Definition 6.1.

Elements of S paired with $1, \ldots, n$			Elements of $T - S$ paired with $n + 1,\;\; n + 2, \ldots$	
1	2 $\;\ldots\;$	n	$n + 1$	$n + 2 \;\ldots$
$\updownarrow$	$\updownarrow \;\ldots\;$	$\updownarrow$	$\updownarrow$	$\updownarrow$
s_1	s_2	s_n	t_1	$t_2 \quad\ldots$

CASE 2. T is finite and S is infinite. This is essentially Case 1 and is left to the reader.

CASE 3. Both S and T infinite. Once again we use the fact that

$$S \cup T = S \cup (T - S).$$

If $T - S$ is finite, then Case 3 can be handled in the same manner as Case 1; the details are left to the reader. Suppose that $T - S$ is infinite. Then S and $T - S$ both contain the same number of elements as any infinite subset of the positive integers. In particular, S contains the same number of elements as M, the set of positive even integers, while $T - S$ contains the same number of elements as W, the set of positive odd integers. Then $S \cup T = S \cup (T - S)$ contains the same number of elements as $W \cup M = N$, the full set of positive integers. Therefore $S \cup T$ is countable.

Corollary. *Let Z be the set of all integers. Then Z is a countable set.*

Proof. The set N^- of negative integers has the same number of elements as the set N of positive integers. This can be shown by pairing the positive integer n with the negative integer $-n$. On the other hand, the set K of integers greater than or equal to zero is also countable (Section 6.1, Exercise 1e). But $Z = N^- \cup K$; that is, the full set of integers is the union of the set of negative integers with the set of integers greater than or equal to zero. Therefore Z is the union of two countable sets, and hence is countable by Proposition 6.

Thus we can say that not only are there as many positive even integers as positive integers, but there are as many positive even integers as there are integers altogether!

The material in Sections 6.2 and 6.3 has aimed for several goals. The first is to give the reader a closer look at the mathematical notion of infinity than he has previously experienced. A second goal, quite closely related to the first, is to show that when we have a good definition to work from, we can do much more than when we deal with intuitive notions. With a purely intuitive notion of the concept of "having the same number," we could not have proceeded very far in any investigation of an infinite set because we could never have "counted" the elements of such a set. By formalizing the notion of two sets having the same number of elements, we have found that we can draw conclusions about infinite sets almost as easily as we can about finite sets. This is not to imply, however, that all the work we did in the previous sections rested on a rigorously formal foundation. Some intuitive concepts were still used, for example, the notions *finite* and *infinite*. Nevertheless, Definition 6.1 helped us to draw a fair number of conclusions about infinite sets.

The reader may have found these conclusions jarring to his intuition; for example, he may have found it repugnant that there is the same number of positive integers as of positive even integers, or that two infinite sets do not necessarily have the same number of elements. Our study of infinity has in fact pointed up the need for that greatest of mathematical virtues, consistency. For if we accept Definition 6.1, we must accept the logical consequences that flow from it. Definition 6.1 was, after all, based on reasonable observations. If we do not like its consequences, we have only two alternatives. First, we can discard the definition and say that although it seemed reasonable when formulated, it simply does not convey the notion it is intended to convey; or second, we can swallow our intuition and accept the definition along with its consequences. But we cannot accept Definition 6.1 and deny its consequences. We must, in other words, be consistent. A definition, or set of axioms, can lead us to conclusions which we might consider quite shocking and strange, but we are not free to deny those conclusions which follow in a logically valid manner from the definition; we can only throw out the definition, if we wish, and seek one that is more to our liking. From experience, however, mathematicians have found that Definition 6.1 is a good definition.

EXERCISES 6.3

1. Prove Proposition 4.

2. Supply the details for Cases 2 and 3 of Proposition 6.

3. Determine whether each of the following statements is true or false. If a statement is true, supply a proof. If a statement is false, correct the statement and prove the corrected statement.

   a) Any two subsets of the set of positive even integers contain the same number of elements.
   b) Any two infinite subsets of the set of integers contain the same number of elements.
   c) If W is any subset of the set Z of positive integers, then infinitely many subsets of Z contain the same number of elements as W. [*Hint:* What if $W = \varnothing$, the empty set?]
   d) An uncountable set may contain a countable subset.
   e) If S is a finite set, then S contains the same number of elements as the collection of subsets of S.

4. A set S is said to have a *greater number of elements* than a set T if T has the same number of elements as some subset W of S, but S and T do not have the same number of elements.

   a) Prove that the set T of Example 4 has a greater number of elements than the set N of positive integers. [*Hint:* We have already shown that N and T do not have the same number of elements, hence all that remains to be shown is that N has the same number of elements as some subset W of T. Pair the positive integer n with the decimal having 1 as its nth digit and 0 for all other digits.]
   b) Prove that the set N of positive integers has a greater number of elements than any finite set has.
   c) Prove that any set which contains at least one element has a greater number of elements than the empty set has.

6.4 THE NOTION OF NUMBER

Definition 6.1 defined only the phrase "has the same number of elements as" and not the concept of number itself. Our purpose in this section will be to try to arrive at a definition of number by means of Definition 6.1. Before venturing a definition of number, however, we shall make a few observations which give some clue as to the form our definition should take.

Imagine a caveman counting his herd of goats. Since our caveman lacks the fine systems of computation that we have today, he would almost certainly use a rather primitive method of counting such as using his fingers and toes. Let us assume that the man has three goats. In counting his goats, the caveman might stick out the thumb of his right hand as he counts one goat, then his right index finger as he counts the second goat, then the right middle finger as he counts the third goat. (This is indeed the way young children often count.) If someone now asks him how many goats he has, he may indicate the number by holding up the thumb, index, and middle fingers of his right hand. Observe that the caveman has indicated the number of goats he has by presenting a set (fingers of his right hand) which has the same number of elements as his herd of goats.

Or perhaps the caveman counts his goats by setting aside a rock for each goat, and when someone inquires as to the number of goats, he shows him the pile of rocks. Again a set with the same number of elements as the herd of goats is used to indicate the number of goats. From a numerical point of view, the caveman's fingers, the pile of rocks, and the herd of goats are the same.

We might then be tempted to say that the number 3 can be considered to be the herd of goats, the three fingers, or the pile of three rocks. That is, we might consider the number 3 actually to be any set with three elements. This is a good start on a notion of the number 3, but we cannot stop yet. The herd of three goats is not the pile of three rocks, thus we cannot say that they are both the number 3, for this would imply that the number 3 is two different things. Actually, to indicate what the number 3 is, any set of three elements is as good as any other. However, in order not to favor one set of three elements above any other set of three elements, and in order to arrive at an absolute definition of the number 3 (absolute in that it is independent of any particular set of three elements), we might consider the number 3 to be *all* sets of three elements.

Definition 6.3. *Let S be any set. Define $\bar{S}$ to be the collection of all sets that have the same number of elements as S. We call $\bar{S}$ the cardinal number of S.**

Example 6. If S represents the caveman's herd of goats, then $\bar{S}$ is the collection of all sets which contain as many elements as the caveman has goats. Thus, $\bar{S}$ is the collection of all sets which have the same number of elements as S. Since S has the same number of elements as itself, S is a member of $\bar{S}$. Let $T = \{1, 2, 3\}$. Then T has the same number of elements as S, hence T is also a member of $\bar{S}$. Since S and T both have

---

* The reader might find it useful at this point to read the discussion following the corollary to Proposition 9 in this section.

the same number of elements, we might suspect that $\overline{S} = \overline{T}$, that is, the cardinal numbers of S and T are the same. Such is indeed the case as we now prove: Suppose W is any set in $\overline{S}$. Then W has the same number of elements as S. Since S has the same number of elements as T, then by Proposition 2, W has the same number of elements as T; hence W is a member of $\overline{T}$. Therefore $\overline{S} \subset \overline{T}$. On the other hand, if V is any member of $\overline{T}$, then (by definition) V has the same number of elements as T. Since S has the same number of elements as T, T has the same number of elements as S (Proposition 1). Therefore by Proposition 2, V has the same number of elements as S; hence V is a member of $\overline{S}$. Consequently, $\overline{T} \subset \overline{S}$. But this means that $\overline{S} = \overline{T}$ (Section 4.3).

It is $\overline{S}$ that we call the cardinal number (or merely the number) 3.

The following propositions give some of the basic properties of cardinal numbers. We should note that some of these properties were already proved in the specialized situation of Example 6.

Proposition 7. *If S is any set, then S is a member of $\overline{S}$, the cardinal number of S.*

Proof. S has the same number of elements as itself.

Proposition 8. *If S has the same number of elements as T, where S and T are both sets, then $\overline{S} = \overline{T}$.*

Proof. To show that $\overline{S} = \overline{T}$, we must show that each member of $\overline{S}$ is a member of $\overline{T}$ and each member of $\overline{T}$ is a member of $\overline{S}$. Suppose first that W is a member of $\overline{S}$. Then (by Definition 6.3) W has the same number of elements as S. Since S has the same number of elements as T, we see (by Proposition 2) that W has the same number of elements as T. Consequently, W is a member of $\overline{T}$; hence $\overline{S} \subset \overline{T}$. Now if V is a member of $\overline{T}$, then V has the same number of elements as T. Since S has the same number of elements as T, T has the same number of elements as S. Therefore V has the same number of elements as T and T has the same number of elements as S, so V has the same number of elements as S; consequently, V is a member of $\overline{S}$. Thus $\overline{T} \subset \overline{S}$, hence $\overline{S} = \overline{T}$.

Example 7. Let N be the set of positive integers. Then $\overline{N}$ is the collection of all infinite countable sets. For if W is any infinite countable set, then W has the same number of elements as N (Proposition 3, Corollary 1), hence W is a member of $\overline{N}$. We thus conclude that any infinite countable set is a member of $\overline{N}$; that is, $\overline{N}$ contains all infinite countable sets. On the other hand, if V is any set which is not an infinite countable set, then either V is finite (in which case V does not have the same number of elements as N) or V is infinite and uncountable in which case again (by definition of "uncountable"), V does not have the same number of elements as N. Therefore $\overline{N}$ contains all infinite countable sets, but no sets which are not infinite and countable. Consequently, $\overline{N}$ consists precisely of the infinite countable sets.

The reader should observe that $\overline{N}$ is an extension of our usual concept of number. We have previously worked only with finite numbers such as 3 and 4, but we might describe $\overline{N}$ as an infinite cardinal number (since it is the cardinal number of an infinite set, *not* because it contains infinitely many members).

Proposition 9. *If a set S does not have the same number of elements as a set T, then $\overline{S}$ and $\overline{T}$ share no members in common; that is, $\overline{S} \cap \overline{T} = \varnothing$. (All we are saying here is, for example, that there is no set which has the same number of elements as a set of three elements and also has the same number of elements as a set of four elements.)*

Proof. Suppose that S and T do not have the same number of elements, but that W is a set in both $\overline{S}$ and $\overline{T}$. Since W is a member of $\overline{S}$, W has the same number of elements as S. Since W is a member of $\overline{T}$, W has the same number of elements as T; consequently, T has the same number of elements as W. Then we see that T has the same number of elements as W and W has the same number of elements as S. Therefore T has the same number of elements as S, contradicting the assumption that S and T do *not* have the same number of elements. The assumption that there was a set W common to $\overline{S}$ and $\overline{T}$ led to a contradiction; therefore there is no set common to both $\overline{S}$ and $\overline{T}$; that is, $\overline{S} \cap \overline{T} = \varnothing$.

Corollary. *If S and T are sets such that $\overline{S} \cap \overline{T} \neq \varnothing$, then S and T have the same number of elements and $\overline{S} = \overline{T}$.*

Proof. Since $\overline{S} \cap \overline{T} \neq \varnothing$, $\overline{S}$ and $\overline{T}$ share some member in common, which by Proposition 9 is impossible if S and T have different numbers of elements. Therefore S and T contain the same number of elements. Then by Proposition 8, $\overline{S} = \overline{T}$.

It is quite possible that the reader finds Definition 6.3 quite disconcerting. First, we are not used to thinking of numbers as collections of sets. To talk about a set being a member of a number may seem totally unappealing. Second, the negative integer -1 is a number, but we have no sets which contain -1 elements, hence there is no cardinal number corresponding to the perfectly decent number -1. Of course, a similar observation holds true for a fractional number such as $\frac{1}{2}$. Third, we have been speaking about a set of three elements as if we already knew what the number 3 was, yet we have not defined it. In other words, is it not some-

what circular to define the cardinal number 3 as the collection of all sets
which contain three elements?

All three of these objections merit attention. In answer to the first—
that we are not used to thinking of a number as a collection of sets—we
might ask, "How exactly *are* we used to thinking of numbers?" It is
probable that the reader's prior concept of number was a highly intuitive
one based mostly on the mechanical procedures of counting, adding, etc.,
that one learns in grammer school. Hopefully, the introductory discussion of this section made Definition 6.3 easier to accept. In Definition 6.3 we have formalized the notion of number and thus put any discussion of number on a more solid foundation than intuition could provide.

As to the second objection—the omission of negative and fractional
numbers from Definition 6.3—the confusion here stems from the fact that
there are actually many different kinds of numbers and not just one.
Cardinal numbers are essentially "counting numbers," or what we might
term "fundamental numbers." (In fact, the term *cardinal* comes from a
Latin word meaning *hinge*, or *pivot*.) Actually, almost any kind of number
can ultimately be defined using the cardinal numbers as a starting point.
Most of the numbers we have considered thus far are the positive integers,
all the integers, and the real numbers; but we have considered all these
kinds in a rather intuitive way. The cardinal numbers are the first numbers we have really tried to define with any rigor. Actually, we can define
the nonnegative integers, that is, the positive integers together with zero,
as the set of cardinal numbers of finite sets.

As to the third point about defining "numbers," we have used the
term "three" right along because from a purely practical viewpoint it is
certain that the reader knows what is meant by a set of three elements.
Computationally, the reader can and does use 3 without first needing a
formal definition of "three." On the other hand, we should note that the
definition of the cardinal number 3 as the collection of all sets having the
same number of elements as $\{1, 2, 3\}$ does not imply or require that we
previously knew what "three" was. The digits 1, 2, and 3 can be considered merely as symbols, without their necessarily referring to some
underlying concept. Likewise, we can define "one" to be the cardinal
number of $\{1\}$ without even knowing what the symbol 1 stands for. We
should also note that *one* can be defined equally well as the cardinal
number of $\{A\}$, or $\{q\}$, or $\{9\}$.

In sum then, our use of terms such as "three" or "number" were an
appeal to the already existing intuitive notions of the reader, an attempt
to motivate what we intended to do formally. Now, in fact, using the
notions of set theory and Definitions 6.1 and 6.3 (together with certain
other definitions we did not introduce), we could develop in a logical and
rigorous fashion the whole theory of numbers as we usually think of them.

We shall not go that far. But in Section 6.5 we shall learn how to add cardinal numbers, and we shall see that the addition of cardinal numbers has precisely the properties we expect of the addition of nonnegative integers.

EXERCISES 6.4

1. Classify the following sets according to cardinal number; that is, group all the sets with the same cardinal number together. Will there be any set which is in two cardinal numbers? Will there be any set which is in no cardinal number?

   a) $\{3, 4\}$
   b) $\{z \mid z$ is a letter of the English alphabet$\}$
   c) $\{x \mid x^2 = -1,$ and x is an integer$\}$
   d) $\{y \mid y^2 = 1,$ and y is an integer$\}$
   e) $\varnothing$
   f) $\{A, B, C, D, E, F\}$
   g) $\{z \mid z$ is less than 1, but greater than 0$\}$
   h) $\{k, g, j\} \cup \{k, m, n\}$
   i) $\{R, S, T\} \cup \{x \mid x$ is an integer between 1 and 23, inclusive$\}$
   j) N, the set of positive integers

2. Consider the cardinal numbers of the sets listed in Exercise 1. Is there any "natural" order in which these cardinal numbers might be arranged? Which would we expect to be larger, the cardinal number of the set N of positive integers or the cardinal number of the set T in Example 4? [Cf. Section 6.3, Example 4.]

3. Prove that any set is a member of exactly one cardinal number.

4. We can define the number 1 to be the cardinal number of $\{A\}$. We know from grammer school that $1 + 1 = 2$. What is the cardinal number of $\{B\}$? What might we define to be the cardinal number of $\{A\} \cup \{B\}$? How might we define the number 3 beginning with the definition of 1?

5. What cardinal number is going to correspond to 0? The answer to this question will become apparent in Section 6.5, but the reader should be able to make an educated guess. [*Hint:* 0 is smaller than 1. What is the only set "smaller" than a set of one element?]

6.5 THE ADDITION OF CARDINAL NUMBERS

For the sake of convenience we shall denote the cardinal number of $\{1\}$ by $\bar{1}$, the cardinal number of $\{1, 2\}$ by $\bar{2}$, and, in general, the cardinal number of $\{1, 2, 3, \ldots, n - 1, n\}$ by $\bar{n}$. Always keep in mind, however, that $\bar{1}$ is *not* $\{1\}$; but rather it is the collection of all sets having the same number of elements as $\{1\}$.

A cardinal number is a collection of sets. Thus, we can speak of a cardinal number α without referring to any particular set which happens to be a member of α. But if α is any cardinal number and A is a member of α, then $\alpha = \overline{A}$. That is, if A is in α, then α is the collection of all sets having the same number of elements as A; thus, by definition, α is $\overline{A}$.

Suppose now that α and $\mathcal{B}$ are any cardinal numbers and we wish to find a natural way to add them. If our definition of addition for cardinal numbers is to be any good, then the definition should give suitable results in cases where we already know what the answer should be. Thus, if we are dealing with the cardinal numbers $\overline{1}$ and $\overline{5}$, we should like to have $\overline{1} + \overline{5} = \overline{6}$.

Taking the union of two sets is very similar to addition and since we are dealing with sets, we might investigate to see whether somehow taking the union of appropriate sets might lead to a suitable definition of the sum of cardinal numbers. For simplicity, we restrict our attention first to the cardinal numbers $\overline{1}$ and $\overline{5}$. Since $\overline{1}$ and $\overline{5}$ are collections of sets, we might try to add them by forming the union $\overline{1} \cup \overline{5}$. We note, however, that $\overline{1} \cup \overline{5}$ is not a cardinal number. A cardinal number contains only sets that have the same number of elements, but $\overline{1} \cup \overline{5}$ contains some sets with only one element, for example, $\{1\}$, and some sets with five elements, for example, $\{1, 2, 3, 4, 5\}$. Since we would expect $\overline{1} + \overline{5}$ to at least be a cardinal number, we must disqualify $\overline{1} \cup \overline{5}$ as a candidate for $\overline{1} + \overline{5}$.

We next observe that $\{1\}$ is a member of $\overline{1}$ and $\{1, 2, 3, 4, 5\}$ is a member of $\overline{5}$. Unfortunately,

$$\{1\} \cup \{1, 2, 3, 4, 5\} = \{1, 2, 3, 4, 5\}$$

is not a member of $\overline{6}$. But $\{A\}$ is also a member of $\overline{1}$, and

$$\{A\} \cup \{1, 2, 3, 4, 5\} = \{A, 1, 2, 3, 4, 5\}$$

is a member of $\overline{6}$; that is, $\{A, 1, 2, 3, 4, 5\}$ does have the same number of elements as $\{1, 2, 3, 4, 5, 6\}$. Thus, the union of suitably chosen members of $\overline{1}$ and $\overline{5}$ does give a member of $\overline{6}$. Observe that the difficulty with $\{1\}$ and $\{1, 2, 3, 4, 5\}$ was that these two sets had an element in common, namely 1, while $\{A\}$ and $\{1, 2, 3, 4, 5\}$ have no elements in common.

Next consider the cardinal numbers $\overline{4}$ and $\overline{3}$. The union of the sets $\{1, 2, 3, 4\}$ and $\{1, 2, 3\}$ from $\overline{4}$ and $\overline{3}$, respectively, is $\{1, 2, 3, 4\}$ which is certainly not a representative of $\overline{7}$, but if we choose sets from $\overline{4}$ and $\overline{3}$ which have no elements in common, for example, if we use the sets $\{a, b, c, d\}$ and $\{1, 2, 3\}$, then

$$\{a, b, c, d\} \cup \{1, 2, 3\} = \{a, b, c, d, 1, 2, 3\}$$

is a member of $\bar{7}$, precisely the cardinal number we would like for the sum of $\bar{3}$ and $\bar{4}$.

We are thus led to the following definition.

Definition 6.4. *Let* α *and* $\mathfrak{B}$ *be any cardinal numbers. We define* $\alpha + \mathfrak{B}$, *the sum of* α *and* $\mathfrak{B}$, *as follows: Select a set S from α and a set T from $\mathfrak{B}$ such that S and T have no elements in common, that is, $S \cap T = \varnothing$. Set $\alpha + \mathfrak{B} = \overline{S \cup T}$, that is, the cardinal number of the union of S and T.*

Definition 6.4 appears to say then that $\bar{1} + \bar{5} = \bar{6}$ and $\bar{4} + \bar{3} = \bar{7}$. But in fact we are not yet sure that Definition 6.4 really does say this. Why? Because as the definition is formulated, we are instructed to choose one member of α and one member of $\mathfrak{B}$ with the stipulation that these members have no elements in common. The reader might be asking himself whether such a choice of members is always possible. Actually the real problem is that there are too many choices of such members. For example, suppose we wish to add the cardinal numbers α and $\mathfrak{B}$ in accordance with Definition 6.4. We can choose any member S' of α and any member T' of $\mathfrak{B}$. We now can get two representatives S and T of α and $\mathfrak{B}$, respectively, such that $S \cap T = \varnothing$ by considering all the elements of S' to be blue and all the elements of T' to be red. This ensures that any member of S will differ from every member of T at least in color; thus S and T have no elements in common. By Definition 6.4, then, $\alpha + \mathfrak{B} = \overline{S \cup T}$. But it may be that there are other members U and V of α and $\mathfrak{B}$, respectively, such that $U \cap V = \varnothing$. If we compute $\alpha + \mathfrak{B}$ using U and V, then we find $\alpha + \mathfrak{B} = \overline{U \cup V}$. If

$$\alpha + \mathfrak{B} = \overline{S \cup T} \qquad \text{and} \qquad \alpha + \mathfrak{B} = \overline{U \cup V},$$

then we surely want $\overline{S \cup T}$ to be the same as $\overline{U \cup V}$; for if $\overline{S \cup T}$ were not the same as $\overline{U \cup V}$, then the sum of α and $\mathfrak{B}$ would depend on which members of α and $\mathfrak{B}$ were used to compute the sum. But if $\alpha + \mathfrak{B}$ is to be the sum of α and $\mathfrak{B}$, it must depend *only* on α and $\mathfrak{B}$, *not* on the representative sets of α and $\mathfrak{B}$ used to compute the sum.

What we are saying is that Definition 6.4 provides us with a rule for computing the sum of two cardinal numbers, namely, for computing $\alpha + \mathfrak{B}$. Suppose one person follows the instructions of Definition 6.4, say to compute $\bar{1} + \bar{5}$, and arrives, as we did, at the answer $\bar{6}$. But suppose someone else also follows the instructions of the definition and, by using members of $\bar{1}$ and $\bar{5}$ different from the ones we used (subject, of course, to the restriction that the members used have no elements in common), arrives at the conclusion that $\bar{1} + \bar{5} = \bar{7}$. If we both followed Definition 6.4 correctly, then we would have to conclude that it is the definition itself which is at fault. That is, the definition would not clearly define what $\bar{1} + \bar{5}$ really is.

We originally used the member $\{A\}$ of $\bar{1}$ and the member $\{1, 2, 3, 4, 5\}$ of $\bar{5}$ to compute $\bar{1} + \bar{5} = \bar{6}$. Suppose we had used the member $\{h\}$ of $\bar{1}$ and the member $\{a, b, c, d, e\}$ of $\bar{5}$ instead. The choice of these members is still in accordance with Definition 6.4 since

$$\{h\} \cap \{a, b, c, d, e\} = \varnothing.$$

And again, as we would hope, $\{h\} \cup \{a, b, c, d, e\}$ is a member of $\bar{6}$. Thus, in two distinct computations we arrive at the conclusion that $\bar{1} + \bar{5} = \bar{6}$.

Now if the sets S and U are both members of the cardinal number $\mathcal{C}$, then S and U have the same number of elements. Similarly, if T and V are each members of the cardinal number $\mathcal{B}$, then T and V each have the same number of elements. As we have seen, if $\mathcal{C} + \mathcal{B}$ when computed with S and T is to be the same as $\mathcal{C} + \mathcal{B}$ when computed with U and V, we must have $\overline{S \cup T} = \overline{U \cup V}$. But the cardinal numbers of $S \cup T$ and $U \cup V$ will be the same if and only if $S \cup T$ and $U \cup V$ have the same number of elements, because Proposition 8 tells us that if $S \cup T$ and $U \cup V$ have the same number of elements, then $\overline{S \cup T} = \overline{U \cup V}$; on the other hand, if $\overline{S \cup T} = \overline{U \cup V}$, then $S \cup T$ and $U \cup V$ are both members of $\overline{S \cup T}$, hence $U \cup V$ has the same number of elements as $S \cup T$. Hence, to show that the sum $\mathcal{C} + \mathcal{B}$ does not depend on the members of $\mathcal{C}$ and $\mathcal{B}$ used to compute the sum, we must prove the following proposition.

Proposition 10. *Suppose $\mathcal{C}$ and $\mathcal{B}$ are any cardinal numbers, S and U are members of $\mathcal{C}$, and T and V are members of $\mathcal{B}$ such that $S \cap T = \varnothing$ and $U \cap V = \varnothing$. Then $S \cup T$ has the same number of elements as $U \cup V$.*

Proof. Since S and U are both members of $\mathcal{C}$, S has the same number of elements as U. And since T and V are both members of $\mathcal{B}$, T has the same number of elements as V. There are therefore pairings π_1 and π_2 between the elements of S and the elements of U, and between the elements of T and the elements of V in accordance with Definition 6.1. A pairing which can be used to show that $S \cup T$ and $U \cup V$ have the same number of elements is illustrated in Fig. 6.3. The reader should supply the details.

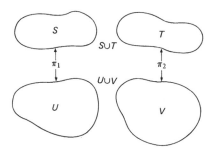

Fig. 6.3

Corollary. *The sum of two cardinal numbers* α *and* ℬ *does not depend on which (nonoverlapping) members of each cardinal number we use to compute the sum.*

We now proceed to prove some of the properties of addition of cardinal numbers.

Proposition 11. *For any two cardinal numbers* α *and* ℬ, α + ℬ = ℬ + α.

Proof. If S and T are members of α and ℬ, respectively, such that

$$S \cap T = \varnothing,$$

then

$$\alpha + \mathcal{B} = (\text{df}) \ \overline{S \cup T} \quad \text{and} \quad \mathcal{B} + \alpha = (\text{df}) \ \overline{T \cup S}.$$

But $S \cup T = T \cup S$ (Chapter 4, Proposition 3). Therefore

$$\overline{S \cup T} = \overline{T \cup S}.$$

Proposition 11, of course, expresses a property which the addition of cardinal numbers must have if it is to be like the ordinary addition of integers.

Proposition 12. *If* α, ℬ, *and* ℭ *are any cardinal numbers, then*

$$\alpha + (\mathcal{B} + \mathcal{C}) = (\alpha + \mathcal{B}) + \mathcal{C}.^*$$

Proof. We may find members S, T, and W of α, ℬ, and ℭ, respectively, such that no two of these sets share any elements in common. Then

$$\alpha + (\mathcal{B} + \mathcal{C}) = \overline{S \cup (T \cup W)} \quad \text{and} \quad (\alpha + \mathcal{B}) + \mathcal{C} = \overline{(S \cup T) \cup W}.$$

But $S \cup (T \cup W) = (S \cup T) \cup W$ (Chapter 4, Proposition 4). Therefore,

$$\alpha + (\mathcal{B} + \mathcal{C}) = \overline{S \cup (T \cup W)} = \overline{(S \cup T) \cup W} = (\alpha + \mathcal{B}) + \mathcal{C}.$$

We now prove that $\overline{\varnothing}$, the cardinal number of the empty set, behaves with respect to addition of cardinal numbers just as 0 does with respect to the addition of integers.

Proposition 13. *If* α *is any cardinal number, then* $\overline{\varnothing} + \alpha = \alpha$.

Proof. Let A be any member of α; $\varnothing$ is the only member of $\overline{\varnothing}$. However, since $\varnothing$ contains no elements whatsoever, A and $\varnothing$ still share no elements

---

* That is, the addition of cardinal numbers is associative. Proposition 11 says that the addition of cardinal numbers is commutative.

in common. Then
$$\overline{\varnothing} + \alpha = (\text{df}) \ \overline{A \cup \varnothing} = \overline{A} = \alpha.$$

[Section 4.4, Exercise 3(e)].

Example 8. We know that addition of cardinal numbers gives the usual result where it should, for example, $\overline{1} + \overline{5} = \overline{6}$. However, there are cardinal numbers for infinite sets as well as for finite sets; hence we can add infinite cardinal numbers too. In particular, we will now compute $\overline{N} + \overline{N}$, where N is the set of positive integers. Since we are adding $\overline{N}$ to $\overline{N}$, we first select any two members S and T of $\overline{N}$, that is, two countable infinite sets, which have no elements in common. Proposition 6 tells us that $S \cup T$ is again a countable set, hence $\overline{S \cup T} = \overline{N}$; that is,

$$\overline{N} + \overline{N} = \overline{N}.$$

Note we could also have chosen S to be the set of positive even integers and T to be the set of positive odd integers. Then $S \cup T = N$, hence again

$$\overline{S \cup T} = \overline{N} + \overline{N} = \overline{N}.$$

The reader may suspect that if we can add cardinal numbers, then we can also multiply them. Such is indeed the case, but we postpone a discussion of this topic until later in the text.

EXERCISES 6.5

1. Supply the details for the proof of Proposition 10.
2. Compute each of the following sums of cardinal numbers by means of the procedure described in Definition 6.4.
   a) $\overline{3} + \overline{1}$ b) $\overline{6} + \overline{2}$ c) $\overline{9} + \overline{11}$
   d) $\overline{3} + \overline{N}$, where N is the set of positive integers   e) $\overline{N} + \overline{7}$
3. We say that a cardinal number α is *less than or equal to* a cardinal number $\mathcal{B}$ if, given some member S of α and some member T of $\mathcal{B}$, then T either has a greater number of elements than S or T has the same number of elements as S (see Section 6.3, Exercise 4). If α is less than or equal to $\mathcal{B}$, we write $\alpha \leq \mathcal{B}$. Prove each of the following.

   a) $\alpha \leq \alpha$ for any cardinal number α.
   b) If α, $\mathcal{B}$, and $\mathcal{C}$ are cardinal numbers such that $\mathcal{B} \leq \mathcal{C}$, then

   $$\alpha + \mathcal{B} \leq \alpha + \mathcal{C}.$$

   c) If α, $\mathcal{B}$, $\mathcal{C}$, and $\mathcal{D}$ are cardinal numbers such that $\alpha \leq \mathcal{B}$ and $\mathcal{C} \leq \mathcal{D}$, then $\alpha + \mathcal{C} \leq \mathcal{B} + \mathcal{D}$.

The following properties could also be proved about $\leq$, but are beyond the scope of this discussion.

d) If α and $\mathcal{B}$ are cardinal numbers such that $\alpha \leq \mathcal{B}$ and $\mathcal{B} \leq \alpha$, then $\alpha = \mathcal{B}$.

e) If α, $\mathcal{B}$, and $\mathcal{C}$ are any cardinal numbers such that $\alpha \leq \mathcal{B}$ and $\mathcal{B} \leq \mathcal{C}$, then $\alpha \leq \mathcal{C}$.

f) Given any two cardinal numbers α and $\mathcal{B}$, either $\alpha \leq \mathcal{B}$ or $\mathcal{B} \leq \alpha$. It should be observed that from our experience with the integers, we would expect $\leq$ to have properties such as (a) through (f).

4. Suppose α and $\mathcal{B}$ are two cardinal numbers. Explain why the following is *not* an appropriate definition of the difference of two cardinal numbers: Select a member S of α and a member T of $\mathcal{B}$. Define $\alpha - \mathcal{B}$ to be $\overline{S - T}$ (cf. Section 4.5). [*Hint:* Does this definition depend on which members of α and $\mathcal{B}$ we use?] Suppose the definition were modified to include the condition that $S \cap T = \varnothing$. Would this make the definition any more suitable? [*Hint:* Does this definition of $\alpha - \mathcal{B}$ give us the properties we might expect subtraction to have?]

chapter seven

The Cartesian Product. Functions

7.1 THE CARTESIAN PRODUCT OF TWO SETS

In the previous chapter we dealt with pairings of the elements of one set with the elements of another set. If π was a pairing of the set S with the set T such that an element s of S was paired with an element t of T in π, then we used the notation

$$s \overset{\pi}{\leftrightarrow} t.$$

This notation, like most mathematical notations, is merely a convenience to help us express certain ideas without an inordinate amount of writing. We might just as well have expressed the idea that s is paired with t in π by writing $s = \pi(t)$, or (s, t) is in π.

Example 1. Suppose $S = \{a, b, c\}$ and $T = \{1, 2, 3\}$. Then one particular pairing π of the elements of S with the elements of T is given by $a \leftrightarrow 1$, $b \leftrightarrow 2$, and $c \leftrightarrow 3$. Using an alternative notation, we might denote the pairing by $(a, 1)$, $(b, 2)$, and $(c, 3)$. If we consider the pairs $(a, 1)$, $(b, 2)$ and $(c, 3)$ as elements, then we can think of the pairing π as a set, namely, $\pi = \{(a, 1), (b, 2), (c, 3)\}$.

Another pairing π' of the elements of S with elements of T is given by $a \leftrightarrow 1$, $b \leftrightarrow 3$, $c \leftrightarrow 3$.* Again we consider π' as a set, namely

$$\pi' = \{(a, 1), (b, 3), (c, 3)\}.$$

In fact, any pairing of the elements of S with the elements of T can be considered as a subset of the set of all pairs of the form (s, t), where s is an element of S and t is an element of T. All possible pairs from these two sets are given in Table 7.1.

---

* This pairing does not satisfy Definition 6.1, but it is nonetheless a valid pairing by some other criterion.

Table 7.1

$(a, 1)$	$(b, 1)$	$(c, 1)$
$(a, 2)$	$(b, 2)$	$(c, 2)$
$(a, 3)$	$(b, 3)$	$(c, 3)$

The pairs in Table 7.1 are *ordered* pairs in the sense that the element of S appears in the first place, while the element of T occurs in the second place. It is the collection of all ordered pairs found in Table 7.1 which gives us what we call the *Cartesian product* of S with T. More generally, we make the following definition.

Definition 7.1. *If S and T are any sets, then the Cartesian product of S with T is defined as the set of all ordered pairs of the form (s, t), where s is an element of S and t is an element of T. We denote the Cartesian product of S with T by $S \times T$.*

Some texts refer to the Cartesian product as merely the *product* of S and T. Another way to read $S \times T$ is S *times* T or S *cross* T.

Example 2. If $S = \{3, 4\}$ and $T = \{r, s, t\}$, then

$$S \times T = (\text{df}) \; \{(3, r), (3, s), (3, t), (4, r), (4, s), (4, t)\},$$

while

$$T \times S = \{(r, 3), (r, 4), (s, 3), (s, 4), (t, 3), (t, 4)\}.$$

We see then that $S \times T$ is not the same set as $T \times S$, but that at least in this example $S \times T$ and $T \times S$ have the same number of elements.

Example 3. Let N be the set of positive integers. Then $N \times N$ is the set of all (ordered) pairs (n, m), where n and m are both positive integers. Typical elements of $N \times N$ are $(9, 8)$, $(5, 101)$, and $(607, 1)$.

Example 4. The faces of an ordinary die are represented by the set

$$S = \{1, 2, 3, 4, 5, 6\}.$$

The set of all possible combinations obtainable from rolling two dice can be represented by the Cartesian product $S \times S$. For example, the element $(3, 4)$ of $S \times S$ represents a 3 on the first die and a 4 on the second die, while $(5, 6)$ represents a 5 on the first die and a 6 on the second.

Definition 7.2. *If (s, t) is an element of $S \times T$, the Cartesian product of the set S with the set T, then s is said to be the first coordinate of (s, t) and t is said to be the second coordinate of (s, t).*

Thus in Example 4, the first coordinate of an element of $S \times S$ represents the roll of the first die, while the second coordinate represents the roll of the second die.

Example 5. The Cartesian plane, sometimes called the coordinate plane, which the reader may know from plane analytic geometry, is nothing more than the Cartesian product of the set R of real numbers with itself, that is, $R \times R$. It is customary to refer to the first coordinate of a point (x, y) of $R \times R$ as the *x-coordinate*, or *abscissa*, and to refer to the second coordinate as the *y-coordinate*, or *ordinate*.

The sets S and T in Example 1 each contain three elements. The set $S \times T$ contains $3 \cdot 3 = 9$ elements. In Example 2 the set S contains two elements, while the set T contains three elements. The set $S \times T$ in this instance contains $2 \cdot 3 = 6$ elements. In general, we have the following proposition.

Proposition 1. *Suppose S is a set which contains m elements and T is a set which contains n elements. Then $S \times T$ is a set which contains $m \cdot n$ elements.*

Table 7.2

$(1, 1)$	$(2, 1)$	$(3, 1)$	$\cdots$	$(m, 1)$
$(1, 2)$	$(2, 2)$	$(3, 2)$	$\cdots$	$(m, 2)$
$\vdots$	$\vdots$	$\vdots$		$\vdots$
$(1, n)$	$(2, n)$	$(3, n)$	$\cdots$	(m, n)

Proof. We let the symbols $1, 2, 3, \ldots, m$ represent the elements of the set S and the symbols $1, 2, 3, \ldots, n$ represent the elements of the set T. The elements of $S \times T$ can be presented in an array such as that shown in Table 7.2. All the elements of $S \times T$ which have 1 as their first coordinate appear in the first column while all the elements of $S \times T$ having 2 as their first coordinate appear in the second column, etc. In the first column there are precisely n elements of $S \times T$ (one for each choice of the second coordinate). In the second column, in fact in each column, there are n elements of $S \times T$. Since there are m columns (one for each choice of the first coordinate), there is a total of $m \cdot n$ elements of $S \times T$.

From Proposition 1 and Example 4, we see that there are 36 possible combinations obtainable from rolling two dice. Of these 36, only one combination, $(1, 1)$, gives us a total of 2 on both dice, while six combinations $(1, 6)$, $(6, 1)$, $(2, 5)$, $(5, 2)$, $(3, 4)$, $(4, 3)$ give us a total of 7.

Suppose now that S, T, and W are any sets. Then $(S \times T) \times W$ is the set of all ordered pairs of the form $((s, t), w)$, where (s, t) is an element of $S \times T$ and w is an element of W. For all intents and purposes, however, the element $((s, t), w)$ is the same as the ordered triple (s, t, w), that is, we have simply left out the inside parentheses. We can therefore define $S \times T \times W$, without writing the parentheses, to be the set of all ordered triples of the form (s, t, w), where s is an element of S, t is an element of T, and w is an element of W.

Example 6. The set of possible combinations obtainable from rolling three dice can be represented by the set $S \times S \times S$, where $S = \{1, 2, 3, 4, 5, 6\}$. There are in all 216 elements of $S \times S \times S$. This becomes clearer if we look at $S \times S \times S$ as $(S \times S) \times S$. We already know that $S \times S$ contains 36 elements. Thus $(S \times S) \times S$ is the Cartesian product of a set containing 36 elements with a set containing 6 elements. Thus, by Proposition 1, $(S \times S) \times S$ contains $36 \cdot 6 = 216$ elements.

Example 7. If R is the set of real numbers, then $R \times R \times R$ is the coordinate space of three dimensions used in solid analytic geometry.

EXERCISES 7.1

1. Write out all the elements of $S \times T$ for each pair of sets S and T given below. It is suggested that the reader use the type of array presented in Tables 7.1 and 7.2.

   a) $S = \{1\}$ and $T = \{a, b, c\}$ b) $S = \{6, 7\}$ and $T = \{a, b, c, d, e\}$

   c) $S = \{-1, 0, 1\}$ and $T = S$ d) $S = \{\%, \$, \#\}$ and $T = \{\cup, \cap, *\}$

2. Prove that $S \times T$ has the same number of elements as $T \times S$ for any sets S and T. Prove that $(S \times T) \times W$ has the same number of elements as $S \times (T \times W)$. Prove that $(S \times T) \times W$ has the same number of elements as $S \times T \times W$.

3. Show that if S is any set, then $S \times \varnothing = \varnothing$; that is, the Cartesian product of any set with the empty set is the empty set.

4. Given: S is a set containing m elements, T is a set containing n elements, and W is a set containing p elements. Prove that $S \times T \times W$ is a set containing $m \cdot n \cdot p$ elements.

5. Prove or disprove each of the following equalities. An equality can be disproved by finding an example for which the equality does not hold. The concept of the equality of two sets was discussed in Section 4.3. It is assumed in each equation that S, T and W are sets.

   a) $(S \times T) \times W = S \times (T \times W)$

   b) $(S \cup T) \times W = (S \times W) \cup (T \times W)$

   c) $(S \cap T) \times W = (S \times W) \cap (T \times W)$

   d) $(S - T) \times W = (S \times W) - (T \times W)$

6. How many combinations are possible from rolling four dice at a time (or from rolling one die four times)? How can the set of combinations be represented? How many combinations are possible when five dice are rolled at a time? n dice? How may each of these sets of combinations be represented?

7.2 THE MULTIPLICATION OF CARDINAL NUMBERS

Proposition 1 indicates that the product of sets is related to the product of numbers. We now use Proposition 1 as the basis for a definition of the product of two cardinal numbers.

Consider the cardinal numbers $\overline{3}$ and $\overline{4}$. A typical member of $\overline{3}$ is $S = \{1, 2, 3\}$, while a typical member of $\overline{4}$ is $T = \{1, 2, 3, 4\}$. Proposition 1 tells us that $S \times T$ contains 12 elements; hence the cardinal number of $S \times T$ is $\overline{12}$, precisely what we should like as the product of $\overline{3}$ and $\overline{4}$. (Note that it is not necessary to have $S \cap T = \varnothing$.) More generally, if S is a member of the cardinal number $\overline{m}$ and T is a member of $\overline{n}$, then Proposition 1 tells us that $S \times T$ will be a member of $\overline{m \cdot n}$. We are thus led to the following definition.

Definition 7.3. *Suppose $\mathfrak{a}$ and $\mathfrak{B}$ are any cardinal numbers. We define the product of $\mathfrak{a}$ and $\mathfrak{B}$ as follows: Select any member S from $\mathfrak{a}$ and any member T from $\mathfrak{B}$. Set $\mathfrak{a} \times \mathfrak{B}$, the product of $\mathfrak{a}$ and $\mathfrak{B}$, equal to $\overline{S \times T}$, the cardinal number of $S \times T$.*

We must now show (as we did when defining the sum of two cardinal numbers) that the product of $\mathfrak{a}$ and $\mathfrak{B}$ does not depend on which members of $\mathfrak{a}$ and $\mathfrak{B}$ are used to compute the product. Specifically, we must prove the following proposition.

Proposition 2. *Suppose $\mathfrak{a}$ and $\mathfrak{B}$ are any cardinal numbers. Let S and U be any members of $\mathfrak{a}$, and let T and V be any members of $\mathfrak{B}$. Then $S \times T$ has the same number of elements as $U \times V$; consequently, $\overline{S \times T} = \overline{U \times V}$. We can therefore say that we arrive at the same result for $\mathfrak{a} \times \mathfrak{B}$ regardless of which members of $\mathfrak{a}$ and $\mathfrak{B}$ are used to compute $\mathfrak{a} \times \mathfrak{B}$.*

Proof. Since S and U are both members of $\mathfrak{a}$, S and U have the same number of elements. Likewise, since T and V are both members of $\mathfrak{B}$, T and V have the same number of elements. Therefore, in accordance with Definition 6.1, there is a pairing π between the elements of S and the elements of U. Similarly, there is a pairing π' between the elements of T and the elements of V proving that T has the same number of elements as V. Let (s, t) be any element of $S \times T$. Since s is an element of S, s is paired with exactly one element u of U in π; that is, $s \overset{\pi}{\leftrightarrow} u$, for some unique element u of U. Likewise, t is paired in π' with exactly one element

v of V. We shall pair the element (s, t) of $S \times T$ with the element (u, v) of $U \times V$. This pairing satisfies the conditions of Definition 6.1. First, we observe that since s is paired only with u (in π) and t is paired only with v, then (s, t) is paired with exactly one element of $U \times V$, namely (u, v). Hence each element of $S \times T$ is paired with exactly one element of $U \times V$. On the other hand, if (u, v) is any element of $U \times V$, then u was paired in π with exactly one element s of S and v was paired in π' with exactly one element t of T. Thus the element (u, v) of $U \times V$ is paired only with the one element (s, t) of $S \times T$. We have therefore shown that $S \times T$ and $U \times V$ contain the same number of elements.

By Proposition 8, Chapter 6, we see that $\overline{S \times T} = \overline{U \times V}$. We therefore conclude that if we use any member of α other than S and any member of $\mathcal{B}$ other than T to compute $\alpha \times \mathcal{B}$, we still arrive at the same result we get by using S and T. Thus we can say that $\alpha \times \mathcal{B}$ is independent of the members of α and $\mathcal{B}$ which are used to compute it.

The following propositions show that multiplication of cardinal numbers has those properties which we expect of simple multiplication.

Proposition 3.   *If α and $\mathcal{B}$ are any cardinal numbers, then $\alpha \times \mathcal{B} = \mathcal{B} \times \alpha$.*

Proof. Suppose S is any member of α and T is any member of $\mathcal{B}$. Then $\alpha \times \mathcal{B} = (\mathrm{df})\ \overline{S \times T}$ and $\mathcal{B} \times \alpha = (\mathrm{df})\ \overline{T \times S}$. But $S \times T$ has the same number of elements as $T \times S$ (Section 7.1, Exercise 2). Consequently,

$$\alpha \times \mathcal{B} = \overline{S \times T} = \overline{T \times S} = \mathcal{B} \times \alpha.$$

Proposition 4.   *If α, $\mathcal{B}$, and $\mathcal{C}$ are any cardinal numbers, then*

$$\alpha \times (\mathcal{B} \times \mathcal{C}) = (\alpha \times \mathcal{B}) \times \mathcal{C}.^*$$

Proof. Let S, T, and W be members of α, $\mathcal{B}$, and $\mathcal{C}$, respectively. We saw in Section 7.1, Exercise 2, that $(S \times T) \times W$ has the same number of elements as $S \times (T \times W)$. But

$$\alpha \times (\mathcal{B} \times \mathcal{C}) = \overline{S \times (T \times W)}$$

and

$$(\alpha \times \mathcal{B}) \times \mathcal{C} = \overline{(S \times T) \times W}.$$

Since $(S \times T) \times W$ has the same number of elements as $S \times (T \times W)$, we have

$$\overline{S \times (T \times W)} = \overline{(S \times T) \times W}.$$

---

* That is, multiplication of cardinal numbers is associative. Proposition 3 says that multiplication of cardinal numbers is commutative. Compare these propositions with Propositions 11 and 12 of Chapter 6.

Consequently,
$$\mathfrak{a} \times (\mathfrak{B} \times \mathfrak{C}) = (\mathfrak{a} \times \mathfrak{B}) \times \mathfrak{C}.$$

Proposition 5. *If $\mathfrak{a}$ is any cardinal number, then $\overline{\varnothing} \times \mathfrak{a} = \overline{\varnothing}$.*

Thus with regard to multiplication as well as addition, $\overline{\varnothing}$ acts like 0.

Proof. Let S be any member of $\mathfrak{a}$. Then $\mathfrak{a} \times \overline{\varnothing} = (\mathrm{df}) \ \overline{S \times \varnothing}$. But $S \times \varnothing = \varnothing$ (Section 7.1, Exercise 3). Therefore

$$\mathfrak{a} \times \overline{\varnothing} = \overline{S \times \varnothing} = \overline{\varnothing}.$$

Addition and multiplication of integers are related by the so-called *distributive law*. That is, if a, b, and c are any integers, then

$$a \cdot (b + c) = a \cdot b + a \cdot c.$$

We now prove that multiplication of cardinal numbers is also distributive over addition of cardinal numbers.

Proposition 6. *If $\mathfrak{a}$, $\mathfrak{B}$, and $\mathfrak{C}$ are any cardinal numbers, then*

$$\mathfrak{a} \times (\mathfrak{B} + \mathfrak{C}) = (\mathfrak{a} \times \mathfrak{B}) + (\mathfrak{a} \times \mathfrak{C}).$$

Proof. Let S, T, and W be members of $\mathfrak{a}$, $\mathfrak{B}$, and $\mathfrak{C}$, respectively, with T and W chosen so that they have no elements in common. Then $T \cup W$ is a member of $\mathfrak{B} + \mathfrak{C}$; therefore $S \times (T \cup W)$ is a member of $\mathfrak{a} \times (\mathfrak{B} + \mathfrak{C})$. The product $S \times T$ is a member of $\mathfrak{a} \times \mathfrak{B}$, and $S \times W$ is a member of $\mathfrak{a} \times \mathfrak{C}$. The sets $S \times T$ and $S \times W$ have no elements in common because no second coordinate of an element of $S \times T$ can be the second coordinate of an element of $S \times W$ (since $T \cap W = \varnothing$). Thus any element of $S \times T$ differs from every element of $S \times W$ at least in the second coordinate. But since $S \times T$ and $S \times W$ have no elements in common,

$$(S \times T) \cup (S \times W)$$

is a member of $(\mathfrak{a} \times \mathfrak{B}) + (\mathfrak{a} \times \mathfrak{C})$.

We now show that $S \times (T \cup W)$ has the same number of elements as $(S \times T) \cup (S \times W)$; moreover, we shall show that they are actually the same set. For if (s, z) is any element of $S \times (T \cup W)$, then s is an element of S, and z is an element of $T \cup W$; that is, z is an element of either T or W. If z is an element of T, then (s, z) is an element of $S \times T$; on the other hand, if z is an element of W, then (s, z) is an element of $S \times W$. In any case, (s, z) is an element of either $S \times T$ or $S \times W$; that is, (s, z) is an element of $(S \times T) \cup (S \times W)$. Thus,

$$S \times (T \cup W) \subset (S \times T) \cup (S \times W).$$

Now if (s, z) is an element of $(S \times T) \cup (S \times W)$, then either (s, z) is an element of $S \times T$, or (s, z) is an element of $S \times W$. In any case, z is an element of either T or W; that is, z is an element of $T \cup W$. Therefore (s, z) is an element of $S \times (T \cup W)$; hence

$$(S \times T) \cup (S \times W) \subset S \times (T \cup W).$$

We therefore have

$$(S \times T) \cup (S \times W) = S \times (T \cup W).$$

But since this is the case, we have

$$\alpha \times (\mathcal{B} + \mathcal{C}) = \overline{S \times (T \cup W)}$$
$$= \overline{(S \times T) \cup (S \times W)} = (\alpha \times \mathcal{B}) + (\alpha \times \mathcal{C}),$$

and the proposition is proved.

Table 7.3

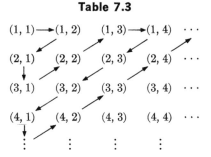

Example 8. Let N be the set of positive integers. In Chapter 6, Example 8, we found that $\overline{N} + \overline{N} = \overline{N}$. What is even more surprising is that $\overline{N} \times \overline{N} = \overline{N}$. We show this by using N itself as a member of $\overline{N}$. Consider the elements of $N \times N$ presented in the array of Table 7.3; that is, all the elements with 1 as the first coordinate are presented in the first row, the elements with 2 as the first coordinate are given in the second row, etc. Pair $(1, 1)$ with 1, $(1, 2)$ with 2, $(2, 1)$ with 3. The procedure is to follow the path suggested by the arrows in Table 7.3 and pair the nth element of $N \times N$ with the positive integer n. Following the path indicated, we ultimately reach any given element of $N \times N$; hence each element of $N \times N$ is paired with precisely one positive integer. But since there are infinitely many elements of $N \times N$, it is also true that each positive integer is paired with precisely one element of $N \times N$. Consequently, there is a pairing of the elements of $N \times N$ with the elements

of N in accordance with Definition 6.1; hence $N \times N$ and N have the same number of elements. This in turn implies that $\overline{N \times N} = \overline{N}$; thus we conclude that $\overline{N} \times \overline{N} = \overline{N \times N} = \overline{N}$.

EXERCISES 7.2

1. Suppose $\mathcal{Q}$, $\mathcal{B}$, and $\mathcal{C}$ are cardinal numbers of finite sets such that $\mathcal{Q} + \mathcal{B} = \mathcal{C} + \mathcal{B}$. Prove that $\mathcal{Q} = \mathcal{C}$.

2. Let $\mathcal{F}$ be the collection of all finite cardinal numbers, that is, cardinal numbers of finite sets. Consider $\mathcal{F} \times \mathcal{F}$, the collection of all ordered pairs of the form $(\mathcal{Q}, \mathcal{B})$, where $\mathcal{Q}$ and $\mathcal{B}$ are finite cardinal numbers. We shall say that a pair $(\mathcal{Q}, \mathcal{B})$ is *equivalent* to the pair $(\mathcal{Q}', \mathcal{B}')$ if $\mathcal{Q} + \mathcal{B}' = \mathcal{Q}' + \mathcal{B}$. Prove each of the following:

   a) $(\mathcal{Q}, \mathcal{Q})$ is equivalent to itself.
   b) If $(\mathcal{Q}, \mathcal{B})$ is equivalent to $(\mathcal{Q}', \mathcal{B}')$, then $(\mathcal{Q}', \mathcal{B}')$ is equivalent to $(\mathcal{Q}, \mathcal{B})$.
   c) If $(\mathcal{Q}, \mathcal{B})$ is equivalent to $(\mathcal{Q}', \mathcal{B}')$ and $(\mathcal{Q}', \mathcal{B}')$ is equivalent to $(\mathcal{Q}'', \mathcal{B}'')$, then $(\mathcal{Q}, \mathcal{B})$ is equivalent to $(\mathcal{Q}'', \mathcal{B}'')$. [*Hint:* Use Exercise 1.]

3. Using the results of Example 8, prove that $\overline{m} \times \overline{N} = \overline{N}$ for any integer m, where N is the set of positive integers.

4. Review the *less than or equal to* definition for cardinal numbers given in Section 6.5, Exercise 3. Prove each of the following.

   a) If $\mathcal{Q} \leq \mathcal{B}$, then $\mathcal{Q} \times \mathcal{C} \leq \mathcal{B} \times \mathcal{C}$, for any cardinal numbers $\mathcal{Q}$, $\mathcal{B}$, and $\mathcal{C}$.
   b) If $\mathcal{Q} \leq \mathcal{B}$, then $\mathcal{Q} \times \mathcal{Q} \leq \mathcal{B} \times \mathcal{B}$, for any cardinal numbers $\mathcal{Q}$ and $\mathcal{B}$.

5. Prove that the Cartesian product of any two countable sets is again a countable set. Prove that if S and T are any nonempty sets such that S is uncountable, then $S \times T$ is uncountable.

7.3 FUNCTIONS

We have frequently used the notion of a pairing of the elements of one set with the elements of another set. If the elements of a set S are being paired with the elements of a set T, then, as we saw in Section 7.1, the pairing can be considered to be a subset of the Cartesian product of S and T.

Some pairings of the elements of one set with the elements of another set satisfy Definition 6.1 and others do not. (See Example 1 in Chapter 7.) If we were pairing the elements of a set S with the elements of a set T, what we might want most is that each element of S be paired with a unique element of T. For example, if we were to form a pairing between the elements of the set $S = \{1, 2, 3\}$ and the elements of the set $T = \{a, b\}$, we would want each element of S paired with some single element of T. We could not satisfy Definition 6.1 using these two sets.

Since the pairing notation we have used up to this point would be somewhat cumbersome in the discussion to follow, we shall switch to the notation of the Cartesian product. We shall also begin to call a pairing by the name it is generally given, *function*. More specifically, we make the following definition.

Definition 7.4. *A subset f of $S \times T$, the Cartesian product of the set S with the set T, is said to be a function from S into T (what we have previously called a pairing of the elements of S with the elements of T) if each element of S appears as a first coordinate exactly once in f. (That is, each element of S is paired with exactly one element of T.)*

Note that the condition that each element of S appear as a first coordinate once in f ensures that each element of S is in fact paired with some element of T; that is, if s is an element of S, then there is some element (s, t) of f (with s as the first coordinate). Then s is paired with t.

Example 9. Let $S = \{1, 2, 3\}$ and $T = \{a, b\}$. We list now all of the functions from S into T:

$$f_1 = \{(1, a), (2, a), (3, a)\}, \qquad f_5 = \{(1, b), (2, a), (3, a)\},$$
$$f_2 = \{(1, a), (2, a), (3, b)\}, \qquad f_6 = \{(1, b), (2, a), (3, b)\},$$
$$f_3 = \{(1, a), (2, b), (3, a)\}, \qquad f_7 = \{(1, b), (2, b), (3, a)\},$$
$$f_4 = \{(1, a), (2, b), (3, b)\}, \qquad f_8 = \{(1, b), (2, b), (3, b)\}.$$

What makes f_1 through f_8 functions is the fact that in each of them, each element of S (that is, 1, 2, and 3) appears as a first coordinate once and only once. Thus

$$\{(1, a), (1, b), (2, a), (3, b)\}$$

would not be a function because the element 1 of S appears as a first coordinate twice. On the other hand,

$$\{(1, a), (2, b)\}$$

is not a function from S into T, since the element 3 of S does not appear as a first coordinate at all.

For comparison, we list below the functions from T into S. Now it is the elements of T which are to appear as the first coordinates.

$$g_1 = \{(a, 1), (b, 1)\}, \qquad g_4 = \{(a, 2), (b, 1)\}, \qquad g_7 = \{(a, 3), (b, 1)\},$$
$$g_2 = \{(a, 1), (b, 2)\}, \qquad g_5 = \{(a, 2), (b, 2)\}, \qquad g_8 = \{(a, 3), (b, 2)\},$$
$$g_3 = \{(a, 1), (b, 3)\}, \qquad g_6 = \{(a, 2), (b, 3)\}, \qquad g_9 = \{(a, 3), (b, 3)\}.$$

Example 10. Let S be the set of hearts and T be the set of people. Define

$f = \{(s, t) \mid s$ is the heart of t, where t is a person and s is a heart$\}$.

Then f is a subset of $S \times T$. Since there are living creatures other than people which have hearts, there are elements of S which are not paired with any element of T; consequently, f is not a function from S into T. On the other hand, if we define

$g = \{(t, s) \mid s$ is the heart of t, where s is a heart and t is a person$\}$,

then since any person has one and only one heart, g is a function from T into S. Note that g, however, is merely the set of elements of f with the coordinates reversed.

It is often inconvenient to write a function from a set S into a set T as a subset of $S \times T$. Although the formal notion of the function as a subset of $S \times T$ is retained, other means of expressing the function are more useful in certain instances. Suppose f is a function from a set S into a set T. If (s, t) is an element of f, then we may write $t = f(s)$. It may be that f is defined by means of some rule or explicit relationship between certain elements of S and certain elements of T. Consider Example 10 in which the rule for g is "s is the heart of t." In such a case, to have the rule is essentially to have the function, and thus the rule is often given and referred to as the function even though the function technically is the subset of $S \times T$ which the rule determines. The following example illustrates this point.

Example 11. Let us suppose that

$$f = \{(n, n^2) \mid n \text{ is a positive integer}\}.$$

Then f is a function from the set N of positive integers into N since each element of N appears as a first coordinate once and only once in f. Looking at it another way, f pairs each positive integer with another positive integer. We see that the rule

$$f(n) = n^2, \text{ where } n \text{ is any positive integer},$$

completely determines f because the rule tells us that f consists of all pairs of the form

$$(n, n^2), \text{ where } n \text{ is any positive integer}.$$

That is, f pairs each positive integer n with the positive integer n^2. It is often more convenient to write

$$f(n) = n^2, \text{ where } n \text{ is any positive integer,}$$

than to write

$$f = \{(n, n^2) \mid n \text{ is a positive integer}\}.$$

Nevertheless, the reader should keep in mind that no matter how we may choose to represent a function from a set S into a set T in any given situation, a function in all cases (according to the way *function* has been defined in Definition 7.4) is actually a subset of $S \times T$. Any notation or device of convenience we use to represent a function must enable us to find this subset unambiguously; if it does not, then the notation does not adequately represent the function.

Example 12. A function cannot be adequately determined by merely the equation $f(x) = x^2$ since this gives no indication of the Cartesian product $S \times T$ of which f is to be a subset. For example, it might be that what we really want to define is

$$f(x) = x^2, \text{ where } x \text{ is any positive integer;}$$

hence f would be a function from the set of positive integers into the set of positive integers. Or it might be that we want

$$f(x) = x^2, \text{ where } x \text{ is any integer at all;}$$

hence f would then be a function from the set of integers into the set of integers.

Observe that in either case f can be considered to be a function from one set S (regardless of whether S is interpreted as the positive integers or as the full set of integers) into the set of nonnegative integers which are perfect squares, or the set of nonnegative integers, or the set of all integers. The point is that if a rule is presented which pairs elements of a set S with those of a set T and the set S is specified while the set T is left indeterminate, then there are many possibilities for T such that the rule determines a function from S into T. If the set T is specified while the set S is left indeterminate, then there may be several sets S such that the rule determines a function from S into T (see Example 12). Thus, strictly speaking, for some rule to define a function from a set S into a set T, the sets S and T must be given, as well as the rule itself.

We observe, however, that if a rule is given which pairs elements of a set S with elements of some set T where T is left indeterminate, then the rule and S determine the same subset of $S \times T$ regardless of what selection of T is made (provided, of course, that T contains all the elements that the given rule pairs with elements of S). For example, if the rule is $f(x) = x^2$, and S is given as the set of positive integers, then regardless of whether T is interpreted as the set of positive integers, the set of positive integers which are perfect squares, the set of nonnegative integers, or the set of all integers, we have

$$f = \{(x, x^2) \mid x \text{ is a positive integer}\}$$

in each case. That is, f always remains the set of all ordered pairs of the form (n, n^2) where n is a positive integer. We could not, however, interpret T as being the set of all houses, because then T would not contain the elements which serve as the second coordinates of the elements of f. The minimum condition here that T must satisfy in order for f to be a function from S into T is that T must contain all squares of positive integers. We further observe that the set of positive integers which are perfect squares is therefore the smallest set compatible with the proposal that f be a function from S into T. The following definition helps us make these reflections more precise.

Definition 7.5. *Let f be a function from a set S into a set T. The set S is said to be the domain of f. The set T is said to be the range of f. The set of all elements of T which appear as second coordinates in f, that is,*

$$\{t \mid (s, t) \text{ is an element of } f\},$$

is said to be the image of f.

There is, of course no need to discuss the set of elements of S which appear as first coordinates in f since, by definition of a function, every element of S must appear as a first coordinate in f.

Example 13. In Example 9, $S = \{1, 2, 3\}$ is the domain of each of the functions f_1 through f_8 while $T = \{a, b\}$ is the range of each of these functions. The image of f_1 is $\{a\}$ and of f_2 is $\{b\}$; the image of f_2 through f_7 is T. The domain of the functions g_1 through g_9 is T, and the range of these functions is S. The image of g_1 is $\{1\}$, of g_5 is $\{2\}$, and of g_9 is $\{3\}$. The image of g_2 and g_4 is $\{1, 2\}$. The computation of the images of g_3, g_6, g_7, and g_8 is left as an exercise.

Example 14. Let us suppose that

$$f = \{(n, n^2) \mid n \text{ is a positive integer}\}.$$

Then f is a function with the set of positive integers as its domain. The image of f is the set of positive integers which are perfect squares. The range of f can be any set which contains the image of f.

We can therefore say that if we know a rule which defines a function and the domain of the function, we essentially know the function, even when the range of the function is not fully specified.

EXERCISES 7.3

1. Find the images of the functions g_3, g_6, g_7, and g_8 of Example 9.

2. In each of the following a rule is given as well as a set S. Decide whether the rule determines a function from the set S into some set T. If the rule does not determine a function, explain why not. If the rule does determine a function, find the smallest set T which can be the range of the function, that is, find the image of the function.

   a) $f(s)$ = the parent of s, where s is an element of S, the set of all people
   b) $f(x) = x^2$, where x is an element of $S = \{12, 13, 15\}$
   c) $g(y) = y - 1$, where y is any element of S, the set of integers
   d) $f(w)$ = the five tallest buildings in w, where w is an element of S, the set of cities of the United States [*Hint:* T will not be a set of buildings in the United States, but T can be a set consisting of sets of five buildings.]
   e) $h(t)$ = the highest mountain in t, where t is an element of S, the set of nations of the world
   f) $g(x)$ = the square root of x, where x is an element of S, the set of positive real numbers
   g) $h(s) = s$, where s is an element of the set S

3. Observe in Example 9 that S contains 3 elements and T contains 2 elements and that there are 2^3 functions from S into T, while there are 3^2 functions T into S. Suppose now that S is a set of n elements and T is a set of 2 elements. Prove that there are 2^n functions from S into T and n^2 functions from T into S. Suppose S is a set of m elements and T is a set of n elements. How many functions do you think there are from S into T? from T into S?

4. Let S be a set of m elements and T be a set of n elements. We can represent the elements of $S \times T$ in an array such as that of Table 7.2. Prove that a subset f of $S \times T$ is a function from S into T if and only if f contains precisely one element from each column of the array. Prove that T is the image of a function f from S into T if and only if f contains at least one element from each row of the array. What can be said about S and T if there is a function f from S into T such that f contains exactly one element from each row and each column of the array?

5. Find all of the functions from $S = \{0, 1\}$ into $T = \{k, m\}$. Find all of the functions from T into S.

6. Prove that a set S has the same number of elements as a set T if and only if there is a function f from S into T such that

$$g = \{(t, s) \mid (s, t) \text{ is an element of } f\}$$

is a function from T into S.

7.4 MORE ABOUT FUNCTIONS

In Example 9 we saw that the functions f_2 through f_8 have the set $T = \{a, b\}$ as both their range and their image. The images of f_1 and f_9, on the other hand, each omit one element of T. Thus the image of a function may or may not be the same as the range of the function. We also saw that the function g_1 pairs two elements of T with the same element of S, while g_2 pairs any two distinct elements of T (actually in this instance T contains only two elements) with two distinct elements of S. The definition of a function f from a set S into a set T demands only that each element of S be used as a first coordinate in f exactly once, but it does not make a similar demand of the elements of T; that is, it is quite possible for some element of T to appear as a second coordinate more than once (just as it is possible for some elements of T not to be used at all as second coordinates in f). Note that all the functions f_1 through f_9 of Example 9 have some element of T appearing at least twice as a second coordinate.

Thus we conclude that a function f from a set S into a set T which either uses each element of T as a second coordinate at least once or uses each element of T as a second coordinate at most once is a special sort of function. This observation leads to the following definition.

Definition 7.6.   A function f from a set S into a set T is said to be onto if the image of f is equal to T, that is, if each element of T appears as a second coordinate in f at least once (or, more informally, if each element of T is paired in f with some element of S). The function f is said to be one-one if no element of T is used as a second coordinate in f more than once (that is, f is one-one if no element of T is paired in f with more than one element of S.)

In Example 9, the functions f_2 through f_7 are onto, while none of the functions f_1 through f_8 are one-one. In that same example, the functions g_2, g_3, g_4, g_6, g_7, and g_8 are one-one, but none of these functions are onto.

In Definition 6.1, we defined the phrase "has the same number of elements as" in terms of a pairing between the elements of one set and the elements of another set. We now convert this definition into function terminology.

Proposition 7.   A set S has the same number of elements as a set T if and only if there is a one-one and onto function from S into T.

Proof.   Suppose S and T have the same number of elements. Then there is a pairing f of the elements of S with the elements of T such that each element of S is paired with precisely one element of T and each element of T is paired with precisely one element of T. Considering f as a subset of $S \times T$ (see Section 7.1), we can represent f as

$$f = \{(s, t) \mid s \text{ is paired with } t \text{ in } f\},$$

thus f is a function from S into T. Since each element of S is paired with precisely one element of T, each element of S then appears as a first coordinate in f once and only once. Since each element of T is paired with precisely one element of S, each element of T appears as a second coordinate in f once and only once. Therefore, as a function, f is both onto and one-one.

Suppose now that f is a function from S into T such that f is both one-one and onto. If we consider an element s of S paired with an element t of T if (s, t) is an element of f, then f pairs each element of S with precisely one element of T. Since f is onto, each element of T appears as a second coordinate in f at least once. On the other hand, since f is one-one, each element of T appears as a second coordinate in f at most once. Consequently, each element of T appears as a second coordinate in f exactly once; that is, each element of T is paired in f with exactly one element of S. We have therefore shown that S has the same number of elements as T.

We proved in Proposition 1 of Chapter 6 that if a set S has the same number of elements as a set T, then T has the same number of elements as S. We shall now reexamine this proof using the function concept. We began with a pairing of the elements of S with the elements of T which satisfied Definition 6.1. That is, in terms of functions, we began with a one-one and onto function f from S into T. We then found a pairing of the elements of T with the elements of S by letting t be paired with s if s was paired with t. In terms of functions, we formed a subset g of $T \times S$ by defining

$$g = \{(t, s) \mid (s, t) \text{ is an element of } f\}.$$

Since f is one-one, each element of t appears as a first coordinate in g at most once (for no element of T appears as a second coordinate more than once in f). On the other hand, since f is onto, each element of T appears as a first coordinate in g at least once (for each element of T appears as a second coordinate in f at least once). Therefore each element of T appears as a first coordinate in g exactly once; consequently, g is a function from T into S. But since each element of S appears as a first coordinate once and only once in f, each element of S appears as a second coordinate once and only once in g. Therefore g is a one-one and onto function from T into S. The function g is merely the function f with the coordinates of the elements reversed. By Proposition 7, the function g suffices to show that T has the same number of elements as S.

If we now let f be any function from a set S into a set T, it is not generally true that $g = \{(t, s) \mid (s, t) \text{ is an element of } f\}$ is a function from T into S; we can see why when we look at the following example.

Example 15. Let N be the set of nonnegative integers and Z be the set of all integers. Then

$$f = \{(z, z^2) \mid z \text{ is an element of } Z\}$$

is a function from Z into N. But

$$g = \{(z^2, z) \mid z \text{ is an element of } Z\}$$

is not a function from N into Z. In the first place, each element of N does not appear as a first coordinate in g; only those elements of N which are perfect squares appear as first coordinates. Hence 0, 1, 4, 9, etc., will appear as first coordinates in g, but integers such as 2, 3, 5, etc., will not. In the second place, we note that all the perfect squares except 0 appear as a first coordinate in two distinct elements of g; for instance, 1 appears as a first coordinate in $(1, 1)$ and $(1, -1)$. Consequently, for two reasons, either one of which would have been sufficient, g is not a function from N into Z.

As part of the paragraph following the proof of Proposition 7, we proved that if f is a one-one and onto function from a set S into a set T, then

$$g = \{(t, s) \mid (s, t) \text{ is an element of } f\}$$

is a function from T into S. We now prove the converse of this result.

Proposition 8. *If f is a function from a set S into a set T such that*

$$g = \{(t, s) \mid (s, t) \text{ is an element of } f\}$$

is a function from T into S, then f is one-one and onto.

Proof. Since g is a function from T into S, each element of t appears as a first coordinate in g once and only once. But then each element of T must appear as a second coordinate in f once and only once; consequently, f is one-one and onto.

Corollary. *A function f from a set S into a set T is one-one and onto if and only if*

$$g = \{(t, s) \mid (s, t) \text{ is an element of } f\}$$

is a function from T into S.

Definition 7.7. *If f is a function from a set S into a set T, then*

$$\{(t, s) \mid (s, t) \text{ is an element of } f\}$$

is called the inverse relation, or merely the inverse, of f. The inverse of f is generally denoted by f^{-1}.

We shall take up *relations* later, but the meaning of Definition 7.7 should be clear even without an explanation of the nature of a relation. The corollary to Proposition 8 tells us that if f is a function from a set S into a set T, then f^{-1} is a function from T into S if and only if f is one-one and onto. If f is any function from a set S into a set T, then the image of f is a subset of T. If we denote the image of f by W, then f can be considered to be an onto function from S into W; that is, by choosing the image of f as the range of f, we make f an onto function. If f happens to be a one-one function, then f^{-1} is a function from W (the image of f) into S; considered as a function from S into W, f is both one-one and onto. We have therefore proved the following result.

Proposition 9. *If f is any one-one function from a set S into a set T and W is the image of f, then f^{-1} is a function from W into S. Moreover, since each element of S appears as a second coordinate once and only once in f^{-1}, we also see that f^{-1} is one-one and onto.*

Example 16. Let f be the function from the set N of positive integers into N defined by $f(n) = n + 1$, for any positive integer n. Then f consists of all ordered pairs of the form $(n, n + 1)$, where n is a positive integer. The function f is one-one. For suppose $f(n) = f(n')$, where n and n' are positive integers; that is, suppose $f(n)$ appears as a second coordinate with both n and n'. Then

$$f(n) = n + 1 = f(n') = n' + 1;$$

hence

$$n + 1 = n' + 1.$$

But substracting 1 from both sides of this equality, we see that $n = n'$. We therefore conclude that $n = n'$ if both $(n, f(n))$ and $(n', f(n))$ are in f. In other words, each element of N is used as a second coordinate at most once in f; therefore f is one-one.

However, the function f is not onto N since 1 does not appear as a second coordinate in f (remember that 0 is not a positive integer). Nevertheless, f is onto if we consider f as a function from N into the image of f, which is

$$W = \{m \mid m \text{ is an integer greater than 1}\}.$$

By Proposition 9,

$$f^{-1} = \{(n + 1, n) \mid n \text{ is a positive integer}\}$$

is a function from W into N. If m is any integer greater than 1, then $f^{-1}(m) = m - 1$; that is

$$f^{-1} = \{(m, m - 1) \mid m \text{ is an integer greater than 1}\}.$$

We might also point out that since f is a one-one and onto function from N into W, f also shows that N has the same number of elements as W.

EXERCISES 7.4

1. Find all of the one-one and onto functions from the set $S = \{1, 2, 3\}$ into itself. There are six such functions.

2. Let Z be the set of integers. Each of the following statements determines a function from Z into Z. Determine whether each of these functions is either one-one or onto. In each case determine the image of the function, and try to find an explicit rule defining the inverse of the function in those cases in which the function is one-one. (See Example 16 for a sample of this type of problem.)

   a) $f(n) = n$, for any integer n
   b) $f(n) = 3$, for any integer n
   c) $f(n) = 2n$, for any integer n
   d) $f(n) = n - 1$, for any integer n
   e) $f(n) = n^3$, for any integer n
   f) $f(n) = n + 9$, for any integer n
   g) $f(n) = \begin{cases} n + 1, & \text{if } n \text{ is greater than } 0, \\ n - 1, & \text{if } n \text{ is less than or equal to } 0 \end{cases}$

3. Let f be a function from the set S into the set T. If A is a subset of T, we define

$$f^{-1}(A) = \{s \mid f(s) \text{ is an element of } A\};$$

   that is $f^{-1}(A)$ is the set of elements s of S such that (s, t) is in f and t is an element of A. Prove each of the following statements about f^{-1}.

   a) If s is an element of $S - f^{-1}(A)$, where A is some subset of T, then $f(s)$ is not an element of A.
   b) If W is the image of f, then $f^{-1}(W) = S$.
   c) If B is any subset of S, we define

$$f(B) = \{f(s) \mid s \text{ is an element of } B\}.$$

   Then $f(f^{-1}(\{t\})) = t$ for any element t in the image of f.
   d) If A is any subset of T, then $f(f^{-1}(A))$ is a subset of A.
   e) If A is any subset of T and f is onto, then $f(f^{-1}(A)) = A$.
   f) $(f^{-1})^{-1} = f$.

4. Let f be a function from the set N of positive integers into N defined by a rule of the form $f(n) = kn + c$, where k and c are fixed positive integers and n is any positive integer; that is, f consists of all ordered pairs of the form $(n, kn + c)$. Prove that any such function f is one-one. Find the image of f in terms of k and c. Prove that f is onto if and only if $k = 1$ and $c = 0$.

5. Let f be a function from the set Z of all integers into Z defined by a rule of the form $f(z) = kz + c$, where k and c are fixed integers and z is any integer. Under what conditions is f one-one? Under what conditions is f onto?

7.5 COMPOSITION OF FUNCTIONS

Let us now reexamine the proof of Proposition 2 of Chapter 6, that is, the proposition which tells us that if a set S has the same number of elements as a set T and T has the same number of elements as a set W, then S has the same number of elements as W. Since S had the same number of elements as T, we first selected a one-one and onto function π from S into T (π was called a pairing in Chapter 6). Since T had the same number of elements as W, we selected a one-one and onto function π' from T into W. From the functions π and π' we obtained a function from S into T in the following manner: For each element s of S, there is a unique element (s, t) of π; and for the element t of T, there is a unique element (t, w) of π'. We then let (s, w) be an element of a new function which we shall denote by g from S into W. That is,

$$g = \{(s, w) \mid \text{for some element } t \text{ of } T,$$
$$(s, t) \text{ is an element of } \pi, \text{ and } (t, w) \text{ is an element of } \pi'\}.$$

We first confirm that g is indeed a function. We must show that each element of S appears as first coordinate precisely once in g. Suppose s is any element of S; then s appears as a first coordinate precisely once in π, say in the element (s, t). Likewise, the element t of T appears as a first coordinate exactly once in π', say in (t, w). Then s appears as a first coordinate only in the element (s, w) of g. Therefore s appears as a first coordinate in the element (s, w) and only in the element (s, w). Each element of S is therefore used as a first coordinate exactly once in g; hence g is a function from S into W.

We next show that g is both one-one and onto (using the fact, of course, that π and π' are both one-one and onto*). Suppose w is any element of W. Since π' is onto and one-one, there is a unique element t of T such that (t, w) is an element of π'. But since π is also one-one and onto, there is a unique element s of S such that (s, t) is an element of π. Then w appears as a second coordinate only in the element (s, w) of g. Consequently, g is one-one and onto. Since we have found a one-one and onto function from S into W, we can use Proposition 7 to conclude that S and W have the same number of elements.

---

* We did not use these facts in showing that g is a function.

Example 17. Let $S = \{1, 2, 3\}$, $T = \{a, b, c\}$, and $W = \{u, v\}$. Suppose that

$$f = \{(1, a), (2, a), (3, b)\} \qquad \text{and} \qquad f' = \{(a, u), (b, v), (c, v)\};$$

then f is a function from S into T and f' is a function from T into W. Note that neither f nor f' is one-one, and f is not onto. Nevertheless, we can still form a function g from S into W as we did in the proof of Proposition 2 of Chapter 6. Specifically, we let

$g = \{(s, w) \mid$ there is an element t of T such that
$\qquad\qquad (s, t)$ is an element of f and (t, w) is an element of $f'\}$.
In this case,

$$g = \{(1, u), (2, u), (3, v)\}.$$

We see that $(1, u)$ is an element of g since $(1, a)$ is in f and (a, u) is in f'; $(2, u)$ is in g since $(2, a)$ is in f and (a, u) is in f'; and $(3, v)$ is in g since $(3, b)$ is in f and (b, v) is in f'. Note also that

$$g(1) = u = f'(f(1)), \qquad g(2) = f'(f(2)), \qquad g(3) = f'(f(3)).$$

We are thus led to make the following definition.

Definition 7.8. *Let f be a function from a set S into a set T and f' be a function from T into the set W. The function*

$\{(s, w) \mid$ *there is an element t of T such that*
$\qquad\qquad (s, t)$ *is an element of f and (t, w) is an element of $f'\}$*

from S into W is called the composition of the function f' with the function f and is denoted by $f' \circ f$.

If f is a function from S into T and f' is a function from T into W, then given any element s of S, $(s, f(s))$ is an element of f (since, by definition, $f(s)$ is the element of T paired with s in f). Because $f(s)$ is an element of T, there is a unique element $(f(s), f'(f(s)))$ in f'. Since $(s, f(s))$ is in f and $(f(s), f'(f(s)))$ is in f', the element $(s, f'(f(s)))$ is an element of $f' \circ f$; in fact, it is the only element with s as a first coordinate. We can therefore write

$$f' \circ f(s) = f'(f(s)).$$

This gives us a specific rule for computing $f' \circ f$.

Example 18. Let f and f' be the functions from N, the set of positive integers, into N defined by $f(n) = n + 1$ and $f'(n) = n^2$ for any positive integer n. Then

$$f' \circ f(n) = f'(f(n)) = f'(n+1) = (n+1)^2.$$

That is, $(n, n+1)$ is an element of f, while $(n+1, (n+1)^2)$ is an element of f', hence $(n, (n+1)^2)$ is an element of $f' \circ f$.

In this instance we can also compute $f \circ f'$. For any positive integer n,

$$f \circ f'(n) = f(f'(n)) = f(n^2) = n^2 + 1.$$

We see then that even when $f' \circ f$ and $f \circ f'$ are both defined, they are not necessarily equal.

If f is a function from S into T and f' is a function from T into W, then $f' \circ f$ always exists, but $f \circ f'$ does not necessarily exist. For if $f \circ f'$ did exist, we would have $f \circ f'(s) = f(f'(s))$ for any element s of S. But $f'(s)$ is defined only when s is an element of T, and $f(f'(s))$ is defined only when $f'(s)$ is an element of S. Since such is not always the case, $f \circ f'$ is not always defined. Of course, $f \circ f'$ does exist in a case like Example 18, in which $S = T = W$.

The following propositions give some of the basic properties of composition of functions.

Proposition 10. *Let f be a function from S into T and g be a function from T into W. Then*

  a) *if f and g are both onto, then $g \circ f$ is onto; and*

  b) *if f and g are both one-one, then $g \circ f$ is one-one.*

Proof of (a). If w is any element of W, then w appears as a second coordinate in at least one element (t, w) of g since g is onto. Since f is onto, t appears as a second coordinate of at least one element (s, t) of f. Therefore w appears as the second coordinate of at least the element (s, w) of $g \circ f$. Since each element of W appears as a second coordinate at least once in $g \circ f$, we know that $g \circ f$ is onto. The proof of (b) is left as an exercise for the reader.

We see in Example 17 that the function $g = f' \circ f$ is onto even though the function f is not onto. We are therefore not entitled to say that $f' \circ f$ is onto if and only if f and f' are both onto.

Proposition 11. *If f is a function from S into T, g is a function from T into W, and h is a function from W into U, then*

$$h \circ (g \circ f) = (h \circ g) \circ f.$$

(That is, the composition of functions is associative. See Proposition 12 of Chapter 6 and Proposition 4 of this chapter.)

Proof. Let s be any element of S. Then

$$h \circ (g \circ f)(s) = h(g \circ f(s)) = h(g(f(s))) = (h \circ g)(f(s)) = (h \circ g) \circ f(s).$$

Proposition 12. *Suppose f is a one-one and onto function from S into T and g is a one-one and onto function from T into W. Then $g \circ f$ is a one-one and onto function from S into W (Proposition 10); consequently, $(g \circ f)^{-1}$ is a function from W into S by the corollary to Proposition 8. In this situation,*

$$(g \circ f)^{-1} = f^{-1} \circ g^{-1}.$$

Proof. By definition, we have

$g \circ f =$ (df) $\{(s, w) \mid$ there is an element t of T such that
                 (s, t) is an element of f and (t, w) is an element of $g\}$.
Therefore

$(g \circ f)^{-1} =$ (df) $\{(w, s) \mid$ there is an element t of T such that
                   (s, t) is an element of f and (t, w) is an element of $g\}$.
Since

$$f^{-1} = \text{(df)} \ \{(t, s) \mid (s, t) \text{ is an element of } f\}$$
and
$$g^{-1} = \text{(df)} \ \{(w, t) \mid (t, w) \text{ is an element of } g\},$$
we have

$(g \circ f)^{-1} =$ (df) $\{(w, s) \mid$ there is an element t of T such that
                   (w, t) is an element of g^{-1} and (t, s) is an element of $f^{-1}\}$.

But this, by definition, is $f^{-1} \circ g^{-1}$. Therefore

$$(g \circ f)^{-1} = f^{-1} \circ g^{-1}.$$

Example 19. Let Z be the set of integers and f and g be the functions from Z into Z defined by $f(z) = z - 1$ and $g(z) = z + 1$ for each integer z. Then f and g are both one-one and onto. Now

$$f^{-1} = \{(z - 1, z) \mid z \text{ is an integer}\} = \{(z, z + 1) \mid z \text{ is an integer}\} = g;$$

similarly, $g^{-1} = f$. That is, $f^{-1}(z) = z + 1$ for any integer z and $g^{-1}(z) = z - 1$ for any integer z. Consequently,

$$(f \circ g)^{-1}(z) = g^{-1}(f^{-1}(z)) = g^{-1}(z + 1) = (z + 1) - 1 = z;$$

similarly, $(g \circ f)^{-1}(z) = z$. Define a function i from Z into Z by $i(z) = z$

for any integer z; that is, $i = \{(z, z) \mid z \text{ is an integer}\}$. Then

$$f \circ g = f \circ f^{-1} = g \circ f = f^{-1} \circ f = i.$$

(Compare this to the $r^{-1} \cdot r = r \cdot r^{-1} = 1$ tor nonzero real numbers.)

EXERCISES 7.5

1. In each of the following exercises we give two functions f and g from $\{a, b, c\}$ into itself. Find $f \circ g$ and $g \circ f$ in each case. In those cases in which f and g are one-one and onto, compute $(f \circ g)^{-1}$ and $(g \circ f)^{-1}$ directly; then compute f^{-1} and g^{-1} and verify that $(f \circ g)^{-1} = g^{-1} \circ f^{-1}$ and $(g \circ f)^{-1} = f^{-1} \circ g^{-1}$.

   a) $f = \{(a, a), (b, a), (c, a)\}$ and $g = \{(a, a), (b, b), (c, a)\}$
   b) $f = \{(a, a), (b, c), (c, b)\}$ and $g = \{(a, b), (b, c), (c, b)\}$
   c) $f = \{(a, b), (b, c), (c, a)\}$ and $g = f$
   d) $f = \{(a, c), (b, c), (c, c)\}$ and $g = \{(a, b), (b, b), (c, b)\}$

2. Let S be any set and denote by $P(S)$ the set of one-one and onto functions from S into S. [A one-one and onto function of a set S into itself is called a *permutation* of S.]

   a) Define the function i from S into S by $i(s) = s$ for each element s of S. Prove that i is an element of $P(S)$.
   b) Suppose f and g are elements of $P(S)$. Prove that $f \circ g$ and f^{-1} are also elements of $P(S)$; that is, show $f \circ g$ and f^{-1} are also one-one and onto.
   c) Suppose f and g are any two elements of $P(S)$. Prove that the equations $f \circ x = g$ and $x \circ f = g$ have a solution in $P(S)$; that is, there are functions h and h' in $P(S)$ such that $f \circ h = g$ and $h' \circ f = g$. [*Hint:* Try $h = f^{-1} \circ g$ and $h' = g \circ f^{-1}$. Compare this with solving an equation like $3x = 4$, or $\frac{1}{2}x = \frac{3}{4}$.]

3. Prove (b) of Proposition 10.

4. Each of the following define one-one and onto functions from the set Z of integers into itself: $f(n) = n + 1$, $g(n) = n - 8$, and $h(n) = n + 2$.

   a) Verify directly that $h \circ (g \circ h) = (h \circ g) \circ h$.
   b) Find g^{-1} and confirm that

   $$g \circ g^{-1}(n) = g^{-1} \circ g(n) = n$$

   for any integer n.
   c) Determine a rule for computing h^{-1} and then find $f \circ h^{-1}$.

5. Let f be a function from a set S into a set T and g be a function from a set T into a set W. What is the domain of $g \circ f$? the range of $g \circ f$? Find an example which shows that the image of $g \circ f$ need not be the same as the image of g. Find an expression for the image of $g \circ f$ in terms of the image of f and the image of g.

chapter eight

Relations

8.1 THE NOTION OF A RELATION

In Chapter 7 we defined a function from a set S into a set T as a special kind of subset of $S \times T$, the Cartesian product of S with T. We considered an element (s, t) of $S \times T$ to be a pairing of the element s of S with the element t of T. The defining characteristic of a function f from S into T is that each element of S appears as a first coordinate exactly once in f. However, we can find important "pairings" of the elements of two sets which are not functions. Let us look at the following examples.

Example 1. If f is a function from a set S into a set T, then f^{-1} is not necessarily a function. Yet f^{-1} is a subset of $T \times S$.

Example 2. Let Z be the set of all integers. We may use the phrase "is less than or equal to" to generate a subset T of $Z \times Z$ in the following manner: We shall let an element (z, z') of $Z \times Z$ be an element of T if and only if z is less than or equal to z'. Since $(1, 0)$ is not an element of T (since 1 is not less than or equal to 0), T is not all of $Z \times Z$. However, each integer appears as a first coordinate in T infinitely many times. For example, 1 appears as a first coordinate in $(1, 1)$, $(1, 2)$, $(1, 3)$, etc. Since the relationship of being less than or equal to is a rather significant one, we might expect the subset T of $Z \times Z$, which is defined by this relationship, to have a certain importance.

The following are properties we associate with "is less than or equal to":

P1. *For any integer z, z is less than or equal to z. (For since z is equal to z, z is less than or equal to z.)*

P2. *If z and z' are integers such that z is less than or equal to z' and z' is less than or equal to z, then $z = z'$.*

P3. *If z, z′ and z″ are integers such that z is less than or equal to z′, and z′ is less than or equal to z″, then z is less than or equal to z″.*

Properties P1 through P3 give us the following properties of the set T.

P1′. *For any integer z, (z, z) is an element of T.* [*Since z is less than or equal to z, by definition of T, the element (z, z) of Z × Z is in T.*]

P2′. *If (z, z′) is an element of T and (z′, z) is also an element of T, then z = z′.*

P3′. *If (z, z′) is an element of T and (z′, z″) is an element of T, then (z, z″) is also an element of T.*

Example 3.   Let S be the set of plane triangles. We may use the phrase "is congruent to" to generate a subset W of $S \times S$ as follows: Let an element $(s, s′)$ be in W if s is congruent to $s′$. The relationship of congruence possesses the following properties:

C1. *Any plane triangle s is congruent to itself.*

C2. *If a triangle s is congruent to a triangle s′, then s′ is congruent to s.*

C3. *If triangle s is congruent to triangle s′, and s′ is congruent to triangle s″, then s is congruent to s″.*

Properties C1 through C3 give the set W the following properties:

C1′. *For any triangle s, (s, s) is an element of W.*

C2′. *If (s, s′) is an element of W, then (s′, s) is also an element of W.*

C3′. *If (s, s′) and (s′, s″) are elements of W, then (s, s″) is an element of W.*

In Examples 2 and 3 we used a relationship between the elements of certain sets to generate, or define, a subset of a Cartesian product. Suppose that instead of starting with a relationship between the elements of a set S and those of a set T, we begin with some subset R of $S \times T$. We could then define a relationship between the elements of S and the elements of T by saying that s is *R-related* to t if (s, t) is an element of R.

Example 4.   Suppose that

$$S = \{1, 2, 3\} \qquad \text{and} \qquad T = \{a, b, c, d\}.$$

Let

$$R = \{(1, a), (1, b), (2, a), (3, d), (4, d)\}.$$

Then 1 would be R-related to both a and b; 2 would be R-related only to a, and 3 and 4 would be R-related only to d.

Example 5. Let

$$S = \{1, 2, 3\} \qquad \text{and} \qquad R = \{(1, 2), (2, 3), (1, 3)\}.$$

Then R is a subset of $S \times S$. The set R has the property that if (s, s') and (s', s'') are in R, then (s, s'') is an element of R. (Compare this statement with P3′ and C3′ in Examples 2 and 3.) That is, if s is R-related to s', and s' is R-related to s'', then s is R-related to s''. Specifically, we have 1 is R-related to 2, 2 is R-related to 3, and 1 is R-related to 3. Observe, however, that (s, s) is not an element of R for any element of S. (Compare this statement with P1′ and C1′.) Since $(1, 2)$ is an element of R, but $(2, 1)$ is not an element of R, it is not true that if s is R-related to s', then s' is R-related to s.

However, it is true that if s is R-related to s', and s' is R-related to s, then $s = s'$. This relationship exists because there are no elements s and s' of S such that (s, s') and (s', s) are both elements of R; in other words, the hypothesis "If s is R-related to s', and s' is R-related to s" is never satisfied in this example; hence the conclusion follows vacuously. (Compare this with the proof following Definition 4.3 that the empty set is a subset of every set.)

A relationship between the elements of one set S and those of some set T can be used to define a subset of $S \times T$. On the other hand, a subset of $S \times T$ can be used to define a relationship between the elements of S and those of T. We might ask which is the better way of considering the relationship between the elements of these two sets—are we to consider it as a subset of $S \times T$ (more precisely, as generating a subset of $S \times T$), or are we to view it as we do relationships in general?

To answer this question, we must make certain distinctions. First, there are what we might call *real* relationships, whose definition and verification pertain to the "real world"; that is, these relationships can be established empirically. For example, "is one mile from," or "is the brother of" are real relationships; one would hardly talk about such relationships in terms of subsets of Cartesian products because there would be little benefit to be gained from such definitions. If T were the subset of $S \times S$ where S is the set of all people (defined by "is the brother of"), then, given T as a starting point rather than the relationship "is the brother of," T itself could be used to define "is the brother of" by defining s to be the brother of s' if and only if (s, s') is an element of T. But there would be little point in defining "is the brother of" by means of a subset of $S \times S$, because the statement "s is the brother of s'" is one which everyone understands and which can be established or disproved without knowing anything about Cartesian products.

On the other hand, there are what might be called *abstract* relationships, such as "less than or equal to," relationships that we consider more intuitively than "is the brother of." For example, if we define the integers by means of a set of axioms, then since we want a "less than or equal to" relationship to pertain to the integers, we must provide for this relationship either directly in the axioms or be able to define "less than or equal to" in terms of the fundamental notions and relationships used in the axioms. In this type of situation, it may be more feasible to deal with a subset of a Cartesian product and to try to put "less than or equal to" on a firm foundation than to risk dealing with an ill-defined and highly informal notion. In sum, when we are dealing with a mathematical relationship which we have heretofore only used in an informal manner, it may be better from a mathematical viewpoint to define a subset of a Cartesian product to be the relation, something which we can then handle rigorously by means of set theory. Let us recall that this approach was very effective in dealing with functions, whereas the more informal notion of a pairing would have proved inadequate for most of the discussion of Chapter 7; or at least, the use of "pairings" would have been considerably more cumbersome.

Both real and abstract, or mathematical, relationships require the set-theory approach for two very good reasons. First, by expressing the properties of any relationship in a set-theoretic way, we can use the impressive machinery of set theory to draw further conclusions about the relationship.

Second, many apparently different relationships are quite similar when viewed as subsets of a Cartesian product. From a set-theoretic viewpoint, the superficially different relationships share many properties in common. Investigating the basic set-theoretic properties of the subset of a Cartesian product generated by some relationship may help us learn more about that relationship, and we may find too that the conclusions are actually valid for a large collection of relationships.

Thus, in the case of certain relationships, the set-theoretic approach is sometimes necessary, or at least desirable, to define the relationship, that is, to put the relationship on a firm axiomatic foundation. In other instances, the set-theoretic approach may help us draw conclusions about certain relationships, or whole groups of relationships, that might be difficult to obtain in any other way. We therefore make the following definition.

Definition 8.1. *If S and T are any sets, then any subset R of $S \times T$ is called a relation between S and T. (A relation between S and T may be any* *subset of* $S \times T$, *including the empty set.*) *If* R *is a relation between* S *and* T *and* (s, t) *is an element of* R, *then* s *is said to be* R-*related to* t. *If* R *is a relation between* S *and* S, *that is, if* R *is a subset of* $S \times S$, *then* R *is said to be a relation on* S.

If we have some relationship between the elements of S and the elements of T, then the relationship defines, or generates, a relation between S and T (see Examples 2 and 3). Similarly, a relation between S and T gives a relationship ("is R-related to") between the elements of S and those of T. Since certain relationships are more important than others, we can reasonably expect that certain types of subsets of $S \times T$ will be more important than others. We shall study several important kinds of relations in the sections following.

EXERCISES 8.1

1. Let $S = \{A, B\}$. Find all of the relations on S (there are 16). Determine which of these relations satisfy the property that they contain (s, s) for each element s of S. Which satisfy the property that they contain (s, s') whenever they contain (s', s)?

2. Suppose R is a relation between S and T and R' is a relation between T and W. Define

$$R^{-1} = \{(t, s) \mid (s, t) \text{ is an element of } R\}$$

and

$$R' \circ R = \{(s, w) \mid \text{there is an element } t \text{ of } T \text{ such that } (s, t) \text{ is an element of } R \text{ and } (t, w) \text{ is an element of } R'\}.$$

Then R^{-1} is a relation between T and S called the *inverse* of R, and $R' \circ R$ is a relation between S and W, called the *composition* of R' with R. (See Definitions 7.7 and 7.8.) Prove each of the following:

a) $(R^{-1})^{-1} = R$, for any relation R between S and T.

b) If R is a relation between S and T, R' is a relation between T and W, and R'' is a relation between W and U, then $R'' \circ (R' \circ R) = (R'' \circ R') \circ R$.

c) If R is a relation between S and T, then $\{(s, s) \mid s \text{ is an element of } S\}$ is a subset of $R^{-1} \circ R$, if given any element s of S, we have sRt for some element t of T.

d) Let $D = \{(s, s) \mid s \text{ is an element of } S\}$ and R be a relation on S. Then $D \circ R = R \circ D = R$.

3. If S is any set, then the *diagonal* D of $S \times S$ is defined as $D = \{(s, s) \mid s \text{ is an element of } S\}$; see Exercise 2(c) and (d). To see why we call D the diagonal, consider the elements of $S \times S$ arrayed as in Table 7.2 and note where the elements of D lie in this array. Let R be a relation on S. In each of the

following we give two statements about R, one "written out" and the other in concise set-theoretic form. In each case, prove that the two statements are equivalent.

a) Each element s of S is R-related to itself. $D \subset R$.
b) If s is R-related to s', then s' is R-related to s. $R^{-1} = R$.
c) If s is R-related to s', and s' is R-related to s, then $s = s'$. $R \cap R^{-1} \subset D$.
d) If s is R-related to s', and s' is R-related to s'', then s is R-related to s''. $R \circ R \subset R$.

4. Let $S = \{a, b, c, d\}$. Each of the following gives a relation R on S. In each case, find R^{-1} and $R \circ R$, and determine which of properties (a) through (d) of Exercise 3 is satisfied by R.

a) $R = \{(a, a), (b, b), (c, c), (d, d), (a, b), (b, d), (a, d)\}$
b) $R = \{(a, b), (c, d), (d, c), (b, a)\}$
c) $R = \{(a, b), (b, c), (c, d), (d, a), (c, b), (a, d)\}$
d) $R = S \times S$
e) $R = \varnothing$

8.2 EQUIVALENCE RELATIONS

A good number of relations have properties quite similar to C1′, C2′ and C3′ of Example 3.

Example 6. Let S be the set of all boards in a certain lumber yard. Let R be the subset of $S \times S$ generated by "has the same length as," that is,

$$R = \{(s, s') \mid s \text{ has the same length as } s'\} \, ;$$

and let R' be the subset of $S \times S$ generated by "is cut from the same wood as." Then R and R' are both relations on S. Although defined differently, R and R' share a number of properties in common:

R1. *For any element s of S, s is R-related to (has the same length as) s.*

R1′. *For any element s of S, s is R'-related to (is cut from the same wood as) s.*

These relations exist because any board is both the same length as itself and cut from the same wood as itself.

R2. *If s is R-related to s', then s' is R-related to s.*

R2′. *If s is R'-related to s', then s' is R'-related to s.*

That is, if s is the same length as s', then s' is the same length as s; and if s is cut from the same wood as s', then s' is cut from the same wood as s.

R3. *If s is R-related to s′ and s′ is R-related to s″, then s is R-related to s″.*

R3′. *If s is R′-related to s′ and s′ is R′-related to s″, then s″ is R′-related to s″.*

That is, if s is the same length as $s′$ and $s′$ is the same length as $s″$, then s is the same length as $s″$; and if s is cut from the same wood as $s′$ and $s′$ is cut from the same wood as $s″$, then we know that s is cut from the same wood as $s″$.

Example 7. Let $\mathfrak{F}$ be the collection of all finite cardinal numbers and consider $\mathfrak{F} \times \mathfrak{F}$. In Section 7.2, Exercise 2, two elements $(\mathcal{C}, \mathcal{B})$ and $(\mathcal{C}′, \mathcal{B}′)$ of $\mathfrak{F} \times \mathfrak{F}$ were defined to be equivalent if $\mathcal{C} + \mathcal{B}′ = \mathcal{C}′ + \mathcal{B}$. The phrase "is equivalent to" generates a relation R on $\mathfrak{F} \times \mathfrak{F}$; that is, "is equivalent to" defines a subset of $(\mathfrak{F} \times \mathfrak{F}) \times (\mathfrak{F} \times \mathfrak{F})$. In parts (a), (b), and (c) of Exercise 2 (Section 7.2), the reader was asked to prove certain properties of "is equivalent to." In terms of the relation R, these properties become:

  a) $(\mathcal{C}, \mathcal{C})$ is R-related to itself.

  b) If $(\mathcal{C}, \mathcal{B})$ is R-related to $(\mathcal{C}′, \mathcal{B}′)$, then $(\mathcal{C}′, \mathcal{B}′)$ is R-related to $(\mathcal{C}, \mathcal{B})$.

  c) If $(\mathcal{C}, \mathcal{B})$ is R-related to $(\mathcal{C}′, \mathcal{B}′)$ and $(\mathcal{C}′, \mathcal{B}′)$ is R-related to $(\mathcal{C}″, \mathcal{B}″)$, then $(\mathcal{C}, \mathcal{B})$ is R-related to $(\mathcal{C}″, \mathcal{B}″)$.

Compare these properties with R1 through R3 and R′1 through R′3 of Example 6.

The relations presented in Examples 3, 6, and 7 all share similar properties. These properties are abstracted in the following definition.

Definition 8.2. *A relation R on a set S is said to be an equivalence relation on S if R has the following properties:*

E1. *Each element s of S is R-related to itself; that is, (s, s) is an element of R for each element s of S.*

E2. *If s is R-related to s′, then s′ is R-related to s; that is, if (s, s′) is an element of R, then (s′, s) is an element of R.*

E3. *If s is R-related to s′ and s′ is R-related to s″, then s is R-related to s″; that is, if (s, s′) is an element of R and (s′, s″) is an element of R, then (s, s″) is also an element of R.*

If R is an equivalence relation S and (s, s′) is an element of R, then we say that s is R-equivalent to s′ or, if no confusion can result, we merely say s is equivalent to s′.

Example 8. In Exercise 2, Section 6.1, and Propositions 1 and 2 of Chapter 6, we essentially proved that the phrase "has the same number of elements as" defines an equivalence relation on the collection of all sets. Two sets having the same number of elements can be thought of as "equivalent" from the standpoint of the number of elements they contain. If R is an equivalence relation on any set S, then the R-equivalent elements of S are usually really equivalent from some particular viewpoint. For instance, in Example 3, two triangles are W-equivalent if they are congruent, that is, if they are geometrically equivalent.

If A is any set, then the cardinal number of A is the collection of all sets, B, such that B has the same number of elements as A. The cardinal number of A is thus the collection of all sets equivalent to A (from the standpoint of number). We proved that each set A is in one and only one cardinal number (Chapter 6, Propositions 7, 8, and 9).

Let us return again to Example 6. Suppose we separate the boards in the lumber yard according to length, stacking all boards of the same length in the same pile. Then each pile of boards will consist of all boards that are R-equivalent to any board in the pile; this is just a fancy way of saying that any board in a given pile will have the same length as any other board in the pile. Furthermore, each board will be in one and only one pile.

If we stack the boards in the lumber yard according to the wood from which they were cut, then again each board will end up in exactly one pile, and each board in any pile will have been cut from the same wood as all other boards in that pile. In this case, a pile of boards will be a set of R'-equivalent boards. If s is any board in the lumber yard, then the pile in which s is found is

$\{b \mid b$ is a board in the lumber yard and b is R'-equivalent to s,

that is, b is cut from the same wood as $s\}$.

Definition 8.3. *If R is an equivalence relation on a set S and s is any element of s, then the set of all elements of S which are R-equivalent to s is called the R-equivalence class, or merely the equivalence class, of s.*

Thus, the cardinal number of a set A is the equivalence class of A with respect to the relation generated on the collection of all sets by the phrase "has the same number of elements as." In Example 6, the R-equivalence class of any board s consists of all boards having the same length as s, while the R'-equivalence class of s consists of all boards cut from the same wood as s. It is not hard to see that in the cases cited in this paragraph, each element of a set is in one and only one equivalence class; for example, each board ends up in one and only one pile whether we stack the boards according to length or according to wood. These observations are made more formal in the following proposition.

Proposition 1. *If R is an equivalence relation on a set S, then each element of S is contained in exactly one R-equivalence class.**

Proof. Let s be any element of S. Then the equivalence class of s is defined as

$$\{w \mid w \text{ is an element of } S \text{ and } w \text{ is } R\text{-equivalent to } s\}.$$

By E1 of Definition 8.2, s is R-equivalent to s; hence s is an element of its own equivalence class. Therefore s is an element of at least one equivalence class. We shall denote the equivalence class of s by $\bar{s}$. We now show that $\bar{s}$ is the only equivalence class which contains s.

Suppose now that s is also an element of the equivalence class $\bar{t}$ of the element t of S. We shall prove that $\bar{s} = \bar{t}$. Let x be any element of $\bar{s}$. Then (x, s) is an element of R. But since s is an element of $\bar{t}$, we also have (s, t) is an element of R. Consequently, both (x, s) and (s, t) are elements of R. By E3 of Definition 8.2, (x, t) is an element of R; hence x is an element of $\bar{t}$. Any element of $\bar{s}$ is therefore an element of $\bar{t}$; that is $\bar{s} \subset \bar{t}$.

Now let x be any element of $\bar{t}$. Then (x, t) is an element of R. Since (s, t) is an element of R, (t, s) is an element of R by E2 of Definition 8.2. Therefore (x, t) and (t, s) are both elements of R; hence (x, s) is an element of R. This means, however, that x is an element of $\bar{s}$. Therefore $\bar{t} \subset \bar{s}$, hence $\bar{s} = \bar{t}$.

Corollary. *If R is an equivalence relation on a set S and $\bar{s}$ denotes the equivalence class of any element s of S, then s is R-equivalent to t if and only if $\bar{s} = \bar{t}$.†*

Proof. Suppose first that $\bar{s} = \bar{t}$. Since s is an element of $\bar{s}$, s is also an element of $\bar{t}$; therefore s is R-equivalent to t.

Next suppose that s is R-equivalent to t. Then s is an element of $\bar{t}$. But s is also an element of $\bar{s}$. By Proposition 1, s is an element of one and only one equivalence class; thus it must be that $\bar{s} = \bar{t}$.

It is hoped that this section has helped demonstrate to the reader the power of abstraction. It took us, for example, quite a bit of Section 6.4 to prove Proposition 1 and its corollary for the one special case when R

---

* That is, R divides, or "partitions," S into nonoverlapping subsets.
† Compare this with Proposition 8 of Chapter 6, where we were dealing with one specific equivalence relation, that is, the relation defined by "has the same number of elements as." Proposition 8 becomes a simple corollary of this more powerful statement pertaining to equivalence relations in general.

is the equivalence relation generated on the collection of all sets by the phrase "has the same number of elements as." Proposition 1 and its corollary are applicable in any situation which involves an equivalence relation. Thus we see that, through the general notion of an equivalence relation, we can study a great many relations at once, and not just one special case at a time.

Cardinal numbers are nothing but the equivalence classes relative to a special equivalence relation. In Sections 6.4 and 7.2 we learned how to add and multiply these equivalence classes and in Section 6.5 we defined an ordering for them. In the Section 8.3, we shall use other equivalence classes to obtain a formal definition of the full set of integers.

EXERCISES 8.2

1. Each of the following parts gives a phrase and a set. Decide in each case whether the phrase generates an equivalence relation on the set. Where the phrase fails to generate an equivalence relation, indicate which properties of an equivalence relation E1, E2, or E3 of Definition 8.2 are lacking.

   a) "has the same mother as"; the set of all people
   b) "has the same area as"; the set of all plane triangles
   c) "is the brother of"; the set of all people
   d) "has twice the area of"; the set of plane triangles
   e) "is less than or equal to"; the set of positive integers
   f) "is similar to"; the set of plane triangles
   g) "leaves the same remainder on division by 2"; the set of all integers

2. Let f be a function from the set S into the set T. If s and s' are any elements of S, we shall say that s is equivalent to s' if $f(s) = f(s')$. Prove that "is equivalent to" generates an equivalence relation on S. Let $\bar{S}$ denote the set of equivalence classes and $\bar{s}$ the equivalence class of any element s of S. Define a function $\bar{f}$ from $\bar{S}$ into T by letting $\bar{f}(\bar{s}) = f(s)$.

   a) Prove that if $\bar{s} = \bar{s}'$, then $\bar{f}(\bar{s}) = \bar{f}(\bar{s}')$, hence the definition of $\bar{f}$ depends only on the equivalence class $\bar{s}$ and not on the element s. [Recall that when we defined the sum of two cardinal numbers we had to be sure that the sum depended only on the cardinal numbers and not on the members of the cardinal numbers used to compute the sum. Here we must be certain that $\bar{f}(\bar{s})$ depends only on $\bar{s}$ and not on the element s.]
   b) Prove that $\bar{f}$ is one-one.

3. Let Z be the set of integers. We say that an integer z is *congruent modulo 3* to an integer z' if z leaves the same remainder on division by 3 as does z'.

   a) Prove that "is congruent modulo 3 to" generates an equivalence relation on Z.
   b) How many equivalence classes are there? [*Hint:* How many possible remainders are there on division by 3?]

   c) An integer which leaves a remainder of 1 on division by 3 can be expressed in the form $3k + 1$, where k is an integer. Prove that the sum of any two integers which leave a remainder of 1 on division by 3 is an integer which leaves a remainder of 2 on division by 3. What can be said about the sum of an integer which leaves a remainder of 2 on division by 3 and another integer of the same type? of an integer which leaves a remainder of 2 on division by 3 and an integer which leaves a remainder of 1?

   d) Let $\bar{0}$ represent the equivalence class of integers which are evenly divisible by 3. What would $\bar{1}$ and $\bar{2}$ represent? Using the observations of (c), how might we define $\bar{1} + \bar{1}$? $\bar{1} + \bar{2}$? Prepare a table to define addition for the equivalence classes $\bar{0}$, $\bar{1}$, and $\bar{2}$.

4. Suppose R and R' are both equivalence relations on some set S. Which properties of an equivalence relation do $R \cup R'$ and $R \cap R'$ necessarily have? You should find, in fact, that $R \cap R'$ is always an equivalence relation.

8.3 A DEFINITION OF THE INTEGERS

Thus far we have considered the set of integers informally; what we said about the integers was based on the probable prior experience of the reader. Such experience with integers was probably largely computational and intuitive; no derivation was made from a solid axiomatic foundation. The purpose of this section is to supply a formal definition of integer which we can use as a basis for a systematic investigation of this type of number.

It should be pointed out, however, that the definition we will consider is only one of a number of rigorous definitions of integer. The reader may legitimately wonder how, given two sets of axioms, we can be sure that they define the same thing. That is, if we are given two formal definitions D and D', each of which purports to define the integers, how can we be sure that what D defines is the same thing as what D' defines? A full discussion of this question is beyond the scope of this text. Nevertheless, we can say that if D implies D', that is, if D' is a theorem in the mathematical system defined by D, and if D' implies D, then for all intents and purposes D and D' are equivalent.

Our starting point will be the collection $\mathfrak{F}$ of cardinal numbers of finite sets. Recall that a set S is finite if S has the same number of elements as a set of the form $\{1, 2, 3, \ldots, n\}$, where $1, 2, 3, \ldots, n$ are considered merely as symbols. If S has the same number of elements as

$$\{1, 2, 3, \ldots, n\},$$

then we denote the cardinal number of S by $\bar{n}$. We already have a definition for the sum and product of any two elements of $\mathfrak{F}$ (Sections 6.5 and 7.2).

[(1,2), (2,3), (3,4), (4,5), . . .] This equivalence class is the integer −1

[(1,1), (2,2), (3,3), (4,4), . . .] This equivalence class is the integer 0

[(2,1), (3,2), (4,3), (5,4), . . .] The integer +1

[(3,1), (4,2), (5,3), (6,4), . . .] The integer +2

Fig. 8.1

Proposition 2. *The sum and product of any two elements of $\mathfrak{F}$ is again an element of $\mathfrak{F}$.*

Proof. Let $\overline{m}$ and $\overline{n}$ be any two elements of $\mathfrak{F}$. The sets $S = \{1, 2, \ldots, m\}$ and $T = \{1', 2', \ldots, n'\}$ may be used as nonoverlapping members of $\overline{m}$ and $\overline{n}$, respectively. Then $\overline{m} + \overline{n} = \overline{S \cup T}$. But $S \cup T$ has the same number of elements as $\{1, 2, \ldots, m, m + 1, m + 2, \ldots, m + n\}$, which is a finite set; hence $\overline{m} + \overline{n} = \overline{S \cup T}$ is an element of $\mathfrak{F}$. Also, $\overline{m} \times \overline{n} = \overline{S \times T}$. But $S \times T$ has the same number of elements as $\{1, 2, 3, \ldots, m \cdot n\}$, hence $S \times T$ is a finite set. Consequently, $\overline{m} \times \overline{n}$ is also an element of $\mathfrak{F}$.

If we think of $\mathfrak{F}$ intuitively as the set of nonnegative integers (with $\overline{\varnothing}$ corresponding to 0), then in order to form the full set of integers, we must somehow bring in the negative integers. In other words, we must find a way to "subtract" any two members of $\mathfrak{F}$.* That is, given $\overline{m}$ and $\overline{n}$ from $\mathfrak{F}$, we must find a suitable way to express $\overline{m} - \overline{n}$.

Let us examine the problem informally for a moment. We may represent the difference between the integers 3 and 2, that is, $3 - 2$, by the pair $(3, 2)$ from $N \times N$, where N is the set of positive integers. Likewise, we might associate $4 - 5$ with the pair $(4, 5)$ and $5 - 4$ with the pair $(5, 4)$. Since $3 - 2$ and $5 - 4$ are both equal to 1, we ought to consider $(3, 2)$ and $(5, 4)$ as being equivalent from the standpoint of the specific integer each represents. (See Fig. 8.1).

Since $6 - 3 = 3$, the integer 3 itself can be associated with the pair $(6, 3)$, or with any one of the pairs $(9, 6)$, $(15, 12)$, or $(4, 1)$. Even though the integer 0 is not a positive integer, we can think of 0 as associated with $(3, 3)$, since $3 - 3 = 0$. Similarly, $(1, 4)$ can be thought of as associated with −3.

Certain pairs of $N \times N$ can be associated with positive integers, other pairs with negative integers, and still other pairs with zero. How can we tell if two pairs (m, n) and (m', n') are to be associated with the same integer? Since (m, n) represents $m - n$ and (m', n') represents $m' - n'$, then (m, n) and (m', n') are associated with the same integer if and only if $m - n = m' - n'$. The catch is, of course, that if we are dealing only

---

* We might even now be able to subtract two suitably chosen elements of $\mathfrak{F}$, but not any two elements.

with the set of positive integers, and addition and multiplication are the only operations we have to work with, then we must express the condition $m - n = m' - n'$ in terms of addition (or multiplication). This, however, is easily taken care of; for by adding $n + n'$ to both sides of

$$m - n = m' - n',$$

we obtain $m + n' = m' + n$. That is,

$$m - n = m' - n' \quad \text{if and only if} \quad m + n' = m' + n.$$

Consequently, we can say that two ordered pairs of positive integers (m, n) and (m', n') are associated with the same integer if and only if

$$m + n' = m' + n.$$

Next, we find a natural way to add and multiply our number pairs. If (m, n) is associated with $m - n$ and (m', n') is associated with $m' - n'$, then we should have $(m, n) + (m', n')$ associated with

$$(m - n) + (m' - n') = (m + m') - (n + n').$$

Since $(m + m') - (n + n')$ is associated with $(m + m', n + n')$, we should define

$$(m, n) + (m', n') = (m + m', n + n').$$

Since

$$(m - n)(m' - n') = m \cdot m' - n \cdot m' - n' \cdot m + n \cdot n'$$
$$= (m \cdot m' + n \cdot n') - (n \cdot m' + n' \cdot m),$$

it follows that we should define $(m, n) \cdot (m', n')$ to be

$$(m \cdot m' + n \cdot n', \, n \cdot m' + n' \cdot m),$$

even though this definition of multiplication appears a bit awkward at first glance.

Our informal discussion concerning pairs of positive integers has been motivation for the following definition.

Definition 8.4. *Let $\mathfrak{F}$ be the set of finite cardinal numbers. We define an element $(\overline{m}, \overline{n})$ of $\mathfrak{F} \times \mathfrak{F}$ to be equivalent to $(\overline{m}', \overline{n}')$ if $\overline{m} + \overline{n}' = \overline{m}' + \overline{n}$. Then "is equivalent to" generates an equivalence relation R on $\mathfrak{F} \times \mathfrak{F}$.* We will denote the set of R-equivalence classes by Z; it is Z we shall call the set of integers. An element of Z will be called an integer. If $(\overline{m}, \overline{n})$ is any*

---

* See Section 7.2, Exercise 2.

element of $\mathfrak{F} \times \mathfrak{F}$, *we shall denote the R-equivalence class of* $(\overline{m}, \overline{n})$ *by* $|\overline{m}, \overline{n}|$. *If* $|\overline{m}, \overline{n}|$ *and* $|\overline{m}', \overline{n}'|$ *are any two integers (remember that, by definition, an integer is an equivalence class), then we define*

$$|\overline{m}, \overline{n}| + |\overline{m}', \overline{n}'| = (\text{df}) \; |\overline{m} + \overline{m}', \overline{n} + \overline{n}'|,$$

and

$$|\overline{m}, \overline{n}| \cdot |\overline{m}', \overline{n}'| = (\text{df}) \; |\overline{m} \times \overline{m}' + \overline{n} \times \overline{n}', \overline{n} \times \overline{m}' + \overline{n}' \times \overline{m}|.$$

If Definition 8.4 seems too much to swallow, keep the following points in mind: We may think of the set of finite cardinal numbers, $\mathfrak{F}$, as being identified with the set of nonnegative integers; moreover, we have defined addition and multiplication on $\mathfrak{F}$ (Sections 6.5 and 7.2). What we lack are the negative integers. Therefore we think of any integer* k as associated with any pair of the form $(\overline{m}, \overline{n})$, where $m - n = k$. The integer k will be associated with both $(\overline{m}, \overline{n})$ and $(\overline{m}', \overline{n}')$ if

$$m - n = m' - n' = k;$$

thus we find that two pairs $(\overline{m}, \overline{n})$ and $(\overline{m}', \overline{n}')$ are equivalent from the standpoint of being associated with the same integer if $\overline{m} + \overline{n}' = \overline{m}' + \overline{n}$. This finding provides the motivation for the equivalence relation R; an R-equivalence class is merely the set of all pairs associated with the same integer. That is, the equivalence class $|\overline{m}, \overline{n}|$ corresponds to the integer $m - n$. Considering $|\overline{m}, \overline{n}|$ as $m - n$ and $|\overline{m}', \overline{n}'|$ as $m' - n'$, we arrive at the definition of multiplication and addition of integers presented in Definition 8.4. We must keep in mind that in motivating Definition 8.4 we have used our intuitive notion of the integers, while Definition 8.4 gives a rigorous definition of the integers based essentially on set theory. Where we said, "$(\overline{m}, \overline{n})$ is associated with the integer $m - n$," it would have been more accurate to say, "$(\overline{m}, \overline{n})$ is associated with what we intuitively think of as the integer $m - n$."

The following proposition shows that addition of integers (here we use "integer" in the strict sense of Definition 8.4) does not depend on which members of the equivalence classes (integers) being added are used to compute the sum.

Proposition 3. *If* $|\overline{m}, \overline{n}| = |\overline{p}, \overline{q}|$ *and* $|\overline{m}', \overline{n}'| = |\overline{p}', \overline{q}'|$, *then*

$$\cdot \; |\overline{m}, \overline{n}| + |\overline{m}', \overline{n}'| = |\overline{p}, \overline{q}| + |\overline{p}', \overline{q}'|.$$

Proof. Since $|\overline{m}, \overline{n}| = |\overline{p}, \overline{q}|$, it follows that $(\overline{m}, \overline{n})$ is equivalent to $(\overline{p}, \overline{q})$; that is, $\overline{m} + \overline{q} = \overline{n} + \overline{p}$. Similarly, $\overline{m}' + \overline{q}' = \overline{n}' + \overline{p}'$. Now

$$|\overline{m}, \overline{n}| + |\overline{m}', \overline{n}'| = (\text{df}) \; |\overline{m} + \overline{m}', \overline{n} + \overline{n}'|$$

---

* "Integer" here is being used in an informal or intuitive way.

and we know, too, that

$$|\overline{p}, \overline{q}| + |\overline{p'}, \overline{q'}| = (\mathrm{df}) \; |\overline{p} + \overline{p'}, \overline{q} + \overline{q'}|.$$

What we must show is that

$$|\overline{m} + \overline{m'}, \overline{n} + \overline{n'}| = |\overline{p} + \overline{p'}, \overline{q} + \overline{q'}|.$$

This will be the case, however, if

$$(\overline{m} + \overline{m'}) + (\overline{q} + \overline{q'}) = (\overline{n} + \overline{n'}) + (\overline{p} + \overline{p'}).$$

Now

$$(\overline{m} + \overline{m'}) + (\overline{q} + \overline{q'}) = (\overline{m} + \overline{q}) + (\overline{m'} + \overline{q'})$$
$$= (\overline{n} + \overline{p}) + (\overline{n'} + \overline{p'}) = (\overline{n} + \overline{n'}) + (\overline{p} + \overline{p'});$$

hence the proposition is proved.

It is left as an exercise to prove that the product of two integers does not depend on the members of the integers which are used to compute the product.

We close this section by *proving* some of the properties of integers that we normally take for granted. Definition 8.4 can, in fact, serve as the basis for a systematic and logically rigorous investigation of the integers, and, ultimately, of the entire real number system. (However, the idea of such an investigation might remind us of the old saw, "Which came first, the chicken or the egg?" In this case we might ask, "If we did not already know what the integers are, how do we know that Definition 8.4 is an adequate definition?" This is a genuinely profound question, and one which troubles many serious mathematicians. Despite our strong intuitive notion of what the integers are, we still need a base to work from in order to prove things about them. What we have done is use our intuitive notion to arrive at a definition of integer, a definition embodying what we hope are enough properties to enable us to derive all the other properties of the integers.)

Proposition 4. *If* $|\overline{m}, \overline{n}|$ *and* $|\overline{p}, \overline{q}|$ *are any integers, then*

$$|\overline{p}, \overline{q}| + |\overline{m}, \overline{n}| = |\overline{m}, \overline{n}| + |\overline{p}, \overline{q}|.$$

Proof

$$|\overline{p}, \overline{q}| + |\overline{m}, \overline{n}| = (\mathrm{df}) \; |\overline{p} + \overline{m}, \overline{q} + \overline{n}| = |\overline{m} + \overline{p}, \overline{n} + \overline{q}|$$
$$(\text{Proposition 11, Chapter 6}) = (\mathrm{df}) \; |\overline{m}, \overline{n}| + |\overline{p}, \overline{q}|.$$

Proposition 5. *If* $|\overline{m}, \overline{n}|$ *is any integer, then*

$$|\overline{\varnothing}, \overline{\varnothing}| + |\overline{m}, \overline{n}| = |\overline{m}, \overline{n}|;$$

moreover, if $|\overline{p}, \overline{q}|$ *is some integer with the property that*

$$|\overline{p}, \overline{q}| + |\overline{m}, \overline{n}| = |\overline{m}, \overline{n}|,$$

for any integer $|\overline{m}, \overline{n}|$, *then*

$$|\overline{\varnothing}, \overline{\varnothing}| = |\overline{p}, \overline{q}|.$$

(*Intuitively, this proposition tells us* $|\overline{\varnothing}, \overline{\varnothing}|$ *acts like zero with respect to addition of integers, and that* $|\overline{\varnothing}, \overline{\varnothing}|$ *is the only integer which acts in this manner.*)

Proof. $|\overline{\varnothing}, \overline{\varnothing}| + |\overline{m}, \overline{n}| =$ (df) $|\overline{\varnothing} + \overline{m}, \overline{\varnothing} + \overline{n}|$. But $\overline{\varnothing} + \overline{m} = \overline{m}$ and $\overline{\varnothing} + \overline{n} = \overline{n}$ (Proposition 13, Chapter 6). Therefore

$$|\overline{\varnothing} + \overline{m}, \overline{\varnothing} + \overline{n}| = |\overline{m}, \overline{n}|;$$

hence $|\overline{\varnothing}, \overline{\varnothing}| + |\overline{m}, \overline{n}| = |\overline{m}, \overline{n}|$.

Now if $|\overline{p}, \overline{q}| + |\overline{m}, \overline{n}| = |\overline{m}, \overline{n}|$ for any integer $|\overline{m}, \overline{n}|$, then

$$|\overline{p}, \overline{q}| + |\overline{\varnothing}, \overline{\varnothing}| = |\overline{\varnothing}, \overline{\varnothing}|.$$

But we also have $|\overline{p}, \overline{q}| + |\overline{\varnothing}, \overline{\varnothing}| = |\overline{\varnothing}, \overline{\varnothing}| + |\overline{p}, \overline{q}| = |\overline{p}, \overline{q}|$. Therefore $|\overline{\varnothing}, \overline{\varnothing}| = |\overline{p}, \overline{q}|$.

Proposition 6. *If* $|\overline{m}, \overline{n}|$ *is any integer, then* $|\overline{m}, \overline{n}| + |\overline{n}, \overline{m}| = |\overline{\varnothing}, \overline{\varnothing}|$. (*That is,* $|\overline{n}, \overline{m}| = -|\overline{m}, \overline{n}|$.)

Proof. $|\overline{m} + \overline{n}, \overline{n} + \overline{m}| = |\overline{\varnothing}, \overline{\varnothing}|$ if $(\overline{m} + \overline{n}) + \overline{\varnothing} = (\overline{n} + \overline{m}) + \overline{\varnothing}$. But

$$(\overline{m} + \overline{n}) + \overline{\varnothing} = \overline{m} + \overline{n} = \overline{n} + \overline{m} \text{ (Proposition 11, Chapter 6)}$$
$$= (\overline{n} + \overline{m}) + \overline{\varnothing};$$

hence the proposition is proved.

EXERCISES 8.3

1. Prove

$$|\overline{3}, \overline{4}| = |\overline{2}, \overline{3}|, \quad |\overline{\varnothing}, \overline{7}| = |\overline{14}, \overline{21}|, \quad |\overline{9}, \overline{11}| = |\overline{\varnothing}, \overline{2}|, \quad |\overline{10}, \overline{6}| = |\overline{4}, \overline{\varnothing}|.$$

2. Prove that the product of two integers does not depend on the members of the integers used to compute the product. Specifically, prove that if

$$|\overline{m}, \overline{n}| = |\overline{p}, \overline{q}| \quad \text{and} \quad |\overline{m}', \overline{n}'| = |\overline{p}', \overline{q}'|,$$

then

$$|\overline{m}, \overline{n}| \cdot |\overline{m}', \overline{n}'| = |\overline{p}, \overline{q}| \cdot |\overline{p}', \overline{q}'|.$$

3. Prove each of the following properties of the integers. In each of the following, $|\overline{m}, \overline{n}|$, $|\overline{p}, \overline{q}|$, and $|\overline{s}, \overline{t}|$ denote arbitrary integers.

   a) $(|\overline{m}, \overline{n}| + |\overline{p}, \overline{q}|) + |\overline{s}, \overline{t}| = |\overline{m}, \overline{n}| + (|\overline{p}, \overline{q}| + |\overline{s}, \overline{t}|)$. (Thus, addition of integers is associative.)

   b) If $|\overline{m}, \overline{n}| + |\overline{s}, \overline{t}| = |\overline{m}, \overline{n}| + |\overline{p}, \overline{q}|$, then $|\overline{s}, \overline{t}| = |\overline{p}, \overline{q}|$. [*Hint:* Cf. Section 7.2, Exercise 1.]

   c) $|\overline{1}, \overline{\varnothing}| \cdot |\overline{m}, \overline{n}| = |\overline{m}, \overline{n}|$.

   d) $|\overline{\varnothing}, \overline{1}| \cdot |\overline{m}, \overline{n}| = |\overline{n}, \overline{m}|$.

   e) $|\overline{\varnothing}, \overline{m}| \cdot |\overline{\varnothing}, \overline{n}| = |\overline{m} \times \overline{n}, \overline{\varnothing}|$. (If we think of integers of the form $|\overline{n}, \overline{\varnothing}|$ as being positive and those of the form $|\overline{\varnothing}, \overline{n}|$ as being negative, then it follows that the product of two negative integers is a positive integer.)

4. Prove that the multiplication of integers is associative.

5. Prove that any element $(\overline{m}, \overline{n})$ of $\mathfrak{F} \times \mathfrak{F}$, where $\mathfrak{F}$ is the set of finite cardinal numbers is equivalent (in the sense of Definition 8.4) to an element in $\mathfrak{F} \times \mathfrak{F}$ of the form $(\overline{\varnothing}, \overline{k})$ or $(\overline{k}, \overline{\varnothing})$, where $\overline{k}$ is an appropriate member of $\mathfrak{F}$. Therefore any integer is of the form $|\overline{\varnothing}, \overline{k}|$, $|\overline{\varnothing}, \overline{\varnothing}|$, or $|\overline{k}, \overline{\varnothing}|$.

6. We shall say that an integer $|\overline{m}, \overline{n}|$ is less than or equal to an integer $|\overline{p}, \overline{q}|$ if $\overline{m} + \overline{q} \leq \overline{n} + \overline{p}$. (See Section 6.5, Exercise 3. You may use any of the results cited in this exercise.) Prove that "less than or equal to" has the following properties.

   a) Any integer is less than or equal to itself.

   b) If $|\overline{m}, \overline{n}|$ is less than or equal to $|\overline{p}, \overline{q}|$, and $|\overline{p}, \overline{q}|$ is less than or equal to $|\overline{m}, \overline{n}|$, then $|\overline{m}, \overline{n}| = |\overline{p}, \overline{q}|$.

   c) If $|\overline{m}, \overline{n}|$ is less than or equal to $|\overline{p}, \overline{q}|$ and $|\overline{p}, \overline{q}|$ is less than or equal to $|\overline{s}, \overline{t}|$, then $|\overline{m}, \overline{n}|$ is less than or equal to $|\overline{s}, \overline{t}|$.

   d) Given any integers $|\overline{m}, \overline{n}|$ and $|\overline{p}, \overline{q}|$, either $|\overline{m}, \overline{n}|$ is less than or equal to $|\overline{p}, \overline{q}|$, or $|\overline{p}, \overline{q}|$ is less than or equal to $|\overline{m}, \overline{n}|$.

   e) If $|\overline{m}, \overline{n}|$ is less than or equal to $|\overline{p}, \overline{q}|$, then $|\overline{m}, \overline{n}| + |\overline{s}, \overline{t}|$ is less than or equal to $|\overline{p}, \overline{q}| + |\overline{s}, \overline{t}|$ for any integer $|\overline{s}, \overline{t}|$.

   f) $|\overline{\varnothing}, \overline{\varnothing}|$ is less than or equal to $|\overline{m}, \overline{n}| \cdot |\overline{m}, \overline{n}|$ for any integer $|\overline{m}, \overline{n}|$.

7. Define a function f from $\mathfrak{F}$ into Z, the set of integers, by $f(\overline{k}) = |\overline{k}, \overline{\varnothing}|$. Prove that f is one-one. Prove that for any $\overline{k}$ and $\overline{k}'$ in $\mathfrak{F}$

$$f(\overline{k} + \overline{k}') = f(\overline{k}) + f(\overline{k}') \qquad \text{and} \qquad f(\overline{k} \times \overline{k}') = f(\overline{k}) \cdot f(\overline{k}').$$

8.4 ORDERINGS

Equivalence relations and functions are not the only types of relations which arise naturally in a wide variety of contexts. In Example 2 we saw that the relation T defined by "is less than or equal to" has the properties P1′, P2′, and P3′. In the following example, we describe another situation in which a relation has similar properties.

Example 9. Let S be some set and $P(S)$ be the collection of all subsets of S. The phrase "is a subset of" defines a relation R on $P(S)$. Specifically, if T and W are any two subsets of S, then (T, S) is an element of R if (and only if) T is a subset of S, that is, $T \subset S$. The relation R has the following properties:

S1. *If W is any subset of S, then (W, W) is an element of R. Since $W \subset W$ (Chapter 4, Proposition 1).*

S2. *If (W, T) and (T, W) are both elements of R, then $W = T$. That is, if $W \subset T$ and $T \subset W$, then $W = T$ (Section 4.3).*

S3. *If (W, T) and (T, U) are both elements of R, then (W, U) is an element of R. That is, if $W \subset T$ and $T \subset U$, then $W \subset U$ (Proposition 2, Chapter 4.)*

Example 10. Let N be the set of positive integers. Then the phrase "evenly divides" defines a relation D on N. The positive integer n evenly divides the positive integer m if m is a multiple of n, that is, if there is a positive integer k such that $m = n \cdot k$. The relation D satisfies the following properties:

D1. *If n is any positive integer, then (n, n) is an element of D. Since $n = n \cdot 1$, n evenly divides itself.*

D2. *If (m, n) and (n, m) are both in D, then $m = n$. Since if m evenly divides n, then m is less than or equal to n, and if n also evenly divides m, we have n is less than or equal to m; hence $m = n$.*

D3. *If (m, n) and (n, p) are in D, then (m, p) is in D. For since m evenly divides n, $n = m \cdot k$; and since n evenly divides p, we have $p = n \cdot k'$. Therefore $p = n \cdot k' = (m \cdot k) \cdot k' = m \cdot (k \cdot k')$. Since $k \cdot k'$ is a positive integer, we have m evenly divides p.*

We abstract the properties of the relations given in Examples 2, 9, and 10 in the following definition.

Definition 8.5. *A relation P on a set S is said to be a partial ordering of S if P satisfies the following conditions:*

P1. *If s is any element of S, then (s, s) is an element of P.*

P2. *If (s, t) and (t, s) are both elements of P, then $s = t$.*

P3. *If (s, t) and (t, w) are both elements of P, then (s, w) is also an element of P.*

A set S with a partial ordering P of S is said to be a partially ordered set.

The reader should note that P1 and P3 are the same as E1 and E3 in Definition 8.2. Thus the only difference between a partial ordering and an equivalence relation is that the property E2 has been replaced by P2. Yet P2 is almost the negation of E2. According to E2, R is a relation on a set S, and when (s, t) is an element of R, then (t, s) is an element of R whereas P2 states that whenever (s, t) is an element of the relation R on the set S, (t, s) is *never* an element of R, except in the instance that $s = t$. Note that if we said that (t, s) was never an element of R if (s, t) is an element of R, we would contradict P1.

The use of the word *ordering* to describe the type of relation defined in Definition 8.5 is not hard to understand. The phrase "less than or equal to" used in Example 2 defines an "ordering" of the integers, real numbers, or any subset thereof, which corresponds with our intuitive concept of "ordering." In the example we are ordering the numbers according to size. Such an ordering is called a "total ordering" since, given any two numbers m and n, either $m \leq n$ or $n \leq m$. The phrase "is a subset of," defining a partial ordering on $P(S)$ in Example 9, also corresponds with our intuitive concept of "ordering"; there, too, we are "ordering" the sets according to "size."

The idea of a *partial* ordering as distinct from a *total* ordering may be entirely new to the reader. Most of the orderings that the reader has encountered are *total*, that is, the ordering somehow relates any two elements being considered. For example, given any two integers m and n, either m is less than or equal to n, or n is less than or equal to m; pick any two integers and you will find that one is less than or equal to the other. However, this is not the case with an arbitrary partial ordering. Let us look at Example 11.

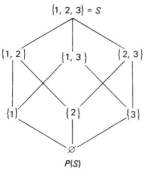

P(S)   **Fig. 8.2**

Example 11. Let $S = \{1, 2, 3\}$, $P(S) =$ the collection of subsets of S, and $R =$ the relation on $P(S)$ defined by "is a subset of." Given any two elements T and W of $P(S)$, it is not necessarily true that either $T \subset W$, or $W \subset T$; that is, there may be two elements of $P(S)$ that are entirely unrelated by R. For example, if $T = \{1, 2\}$ and $W = \{2, 3\}$, then T is not a subset of W, nor is W a subset of T. (See Fig. 8.2, a diagram of the relationships among the subsets of S relative to $\subset$.)

The reader should beware that some relations which appear at first glance to be partial orderings are really not.

Example 12. Let S be the set of plane triangles. A relation R is defined on S by the phrase "has area less than or equal to the area of"; that is, (s, t) is an element of R, where s and t are plane triangles, if s has area less than or equal to the area of t. It is easily verified that R has properties P1 and P3; however, R does not have P2 for it is possible that, given two *distinct* triangles s and t, the area of s is less than or equal to the area of t, and the area of t is less than or equal to the area of s. (All we can conclude is that s and t have the same area, but not that $s = t$.)

In some orderings, such as that given in Example 2, any two elements of the partially ordered set are related in some way by the ordering. In other orderings, two elements of the partially ordered set may not be related in any way by the ordering (see Example 11). Some orderings are truly *partial*, while some are *total*. Thus we make the following definition.

Definition 8.6. *A partial ordering P on a set S is said to be a total ordering if, given any elements s and t of S, either (s, t) is an element of P or (t, s) is an element of P; that is, for any elements s and t of S, either s is P-related to t or t is P-related to s.*

The ordering defined on the set of integers by "less than or equal to" is a *total ordering*. It should, of course, always be kept in mind that a total ordering is just a special case of a partial ordering, and, consequently, has properties P1, P2, and P3.

The notion of order brings to mind such words as *least, greatest, successor*, and *predecessor*, to mention but a few.

Example 13. Let S, $P(S)$, and R be as in Example 11. The empty set $\varnothing$ is a subset of any set. Thus we can say that $\varnothing \subset T$ for any subset T of S. Consequently, $(\varnothing, T)$ is an element of R for any subset T of S. On the other hand, if T is any subset of S, $T \subset S$ (this is a tautology); hence (T, S) is an element of R for any member T of $P(S)$. We may think of $\varnothing$ as being the *least* element of $P(S)$, and S as being the *greatest* element of R; or, we may say that $\varnothing$ is a *minimal* element of $P(S)$, while S is a *maximal* element. Each element of $P(S)$ is R-related to S, while $\varnothing$ is R-related to each element of $P(S)$.

Let $T = \{1\}$ and $W = \{2\}$. Then neither (T, W) nor (W, T) is an element of R. We know that (T, S) and (W, S) are both elements of R. There is, however, a subset of S "smaller" than S, namely, $\{1, 2\}$, such that $(T, \{1, 2\})$ and $(W, \{1, 2\})$ are both elements of R. However, with this property, $\{1, 2\}$ is in a sense the *least* subset of S. We might also say that $\varnothing$ is the *greatest* subset of S with the property that $(\varnothing, T)$ and $(\varnothing, W)$ are both elements of R.

Example 14. Let Z be the set of integers and R be the relation defined on Z by the phrase "is less than or equal to." Take $W = \{1, 2, 3, 4\}$ and consider the integer 6. Then 6 has the property that $(1, 6)$, $(2, 6,)$, $(3, 6)$, and $(4, 6)$ are all elements of R; in fact, $(w, 6)$ is an element of R for any element w of W. Any integer greater than 4 also has this property. But we note that for 4 as well, $(1, 4)$, $(2, 4)$, $(3, 4)$, and $(4, 4)$ are also all elements of R. If n is any integer with the property that (w, n) is an element of R for any element w of W, then since 4 is an element of W, it follows that $(4, n)$ is an element of R. We therefore conclude that 4 is the least element of Z with the property that (w, n) is an element of R for any element w of W. Similar reasoning shows that 1 is the greatest integer n such that (n, w) is an element of R for any element w of W.

We make the ideas presented in Examples 13 and 14 more precise in the following definition.

Definition 8.7. *Let S be a set partially ordered by the relation P. Let W be any nonempty subset of S. An element u of S is said to be an upper bound for W if (w, u) is an element of P, that is, if w is P-related to u, for any element w of W. (Intuitively, we know that an upper bound for W is an element of S which is "larger"—with respect to the ordering being used—than any element of W.) An element v of S is said to be a lower bound for W if (v, w) is an element of P, that is, if v is P-related to w, for any element w of W. (Intuitively, we know that v is "smaller" than any element of W.)*

An upper bound u of W is called a least upper bound for W if u (1) is an upper bound and (2) is P-related to any upper bound of W. Similarly, v is a greatest lower bound for W if (1) v is a lower bound for W, and (2) any lower bound for W is P-related to v. (Intuitively, we know that a least upper bound is the "smallest" element which is larger than all the elements of W, and that a greatest lower bound is the "largest" element which is smaller than all the elements of W.)

We see in Example 14 that 6 is an upper bound for $\{1, 2, 3, 4\}$, but 4 is the least upper bound; 1 is the greatest lower bound of $\{1, 2, 3, 4\}$. In Example 13, S is an upper bound for $\{W, T\}$, but $\{1, 2\}$ is the least upper bound for $\{W, T\}$. The only lower bound as well as the greatest lower bound for $\{W, T\}$ is $\varnothing$.

Example 15. Let N be the set of positive integers partially ordered by the relation P defined by "less than or equal to." The set N has no upper bound, that is, there is no positive integer which is larger than all other positive integers. The positive integer 1 is the greatest and only lower bound for N; that is, 1 is the only positive integer which is less than or equal to every positive integer.

The upper and lower bounds for a subset of a partially ordered set depend on the partial ordering. This point is illustrated in the following example.

Example 16. Let N be the set of positive integers partially ordered by the relation D discussed in Example 10. Let $W = \{3, 4\}$. Then a positive integer u is an upper bound for W if 3 evenly divides u, and 4 evenly divides u; that is, if $(3, u)$ and $(4, u)$ are both elements of D. The integers 12, 24, 36, 48, etc., are all upper bounds for W, and 12 is a least upper bound. Note, however, that 12 is a least upper bound for W *not* because 12 is less than any other upper bound but because 12 evenly divides every upper bound; that is, if u' is an upper bound for $\{3, 4\}$, then 12 evenly divides u'. Since 1 is the only positive integer which divides both 3 and 4, 1 is the greatest lower bound of W.

With respect to the partial ordering of N defined by "is less than or equal to," the least upper bound and greatest lower bound of W are 4 and 3, respectively.

Example 17. Let $P(S)$ be the collection of all subsets of some set S, and R be the partial ordering defined on S by "is a subset of" (Example 9). Suppose A is a collection of subsets of S. Then U is an upper bound for A if each set in A is a subset of U, and U is a least upper bound for A if U is an upper bound for A and, in turn, is a subset of any upper bound for A. That is, U is an upper bound for A if each member of A is a subset of U, and U itself is a subset of any set of which each member of A is a subset. If $A = \{W, T\}$, it is not hard to show that the least upper bound for A is $W \cup T$ and the greatest lower bound for A is $S \cap T$.

We have generally referred to *the* least upper bound, or *the* greatest lower bound for a set. A subset of a partially ordered set may have neither a least upper bound, nor a greatest lower bound (for example, the ordered set of integers has neither an upper nor a lower bound), but if the set has a least upper bound (greatest lower bound), that bound is unique, as we now prove.

Proposition 7. *Suppose S is a set partially ordered by the relation R, W is a subset of S, and u and u' are least upper bounds for W. Then $u = u'$.*

Proof. Since u is a least upper bound for W and u' is an upper bound for W, then, by Definition 8.7, u is R-related to u'. On the other hand, since u' is a least upper bound for W and u is an upper bound for W, it follows that u' is R-related to u. Since u is R-related to u' and u' is R-related to u, we know by P2 that $u = u'$.

EXERCISES 8.4

1. Let Z be the set of integers. Define a relation R on Z by defining (n, m) to be an element of R if $n = m$ or $n < m - 1$ (but $n \neq m - 1$), where n and m are integers.

    a) Prove that R is a partial ordering of Z.

    b) Prove that R is not a total ordering of Z. [*Hint:* Prove that 3 and 4 are not related in any way by R.]

    c) Prove that $\{3, 4\}$ has no least upper bound. [*Hint:* Prove that 6 and 7 are both upper bounds for $\{3, 4\}$, but 6 is not R-related to 7.]

    d) Let M be the set of even integers. Prove that M is a totally ordered subset of Z. That is, show that for any two even integers k and k', either k is R-related to k' or k' is R-related to k.

2. Prove in Example 17 that $W \cup T$ is the least upper bound for $\{W, T\}$ and $T \cap W$ is the greatest lower bound for $\{W, T\}$.

3. Suppose S is a set partially ordered by the relation R, W is a subset of S, and v and v' are greatest lower bounds for W. Prove $v = v'$.

4. Find the least upper bound and greatest lower bound for each of the following subsets of the set N of positive integers partially ordered by D as in Example 10:

    a) $\{4, 5\}$ b) $\{2, 3\}$ c) $\{6, 4\}$

    d) $\{2, 3, 4\}$ e) $\{3, 6, 9\}$ f) $\{6, 10, 14\}$

    The least upper bound of a subset $\{n, m\}$ is called the *least common multiple* of m and n, and the greatest lower bound of $\{m, n\}$ is called the *greatest common divisor* of m and n.

5. Let S be a set partially ordered by the relation R, and T be a set partially ordered by the relation R'. We define a relation R'' on $S \times T$, the Cartesian product of S with T, as follows: Let (s, t) be R''-related to (s', t') if s is R-related to s' and t is R'-related to t'.

    a) Prove that R'' is a partial ordering on $S \times T$. *Illustration:* Let S and T both be the set of integers and R and R' both be the relation defined by "is less than or equal to." Then (m, n) is R''-related to (m', n') if m is less than or equal to m' and n is less than or equal to n'.

b) Suppose $W \subset S$ and $Z \subset T$ with u as the least upper bound of W and u' as the least upper bound of T. Prove that (u, u') is the least upper bound of $W \times Z$ in $S \times T$ (with respect to the partial ordering R'').

6. Let S be a set partially ordered by the relation R. We say that S is a *lattice* if any subset of S which consists of two elements has both a least upper bound and a greatest lower bound. If $\{s, s'\}$ is any two-element subset of S (with the possibility of $s = s'$), we shall denote the least upper bound of $\{s, s'\}$ by $s \vee s'$ and the greatest lower bound of $\{s, s'\}$ by $s \wedge s'$. Suppose that S with partial ordering R is a lattice; prove each of the following:

a) $s \vee s = s$ for any element s of S.
b) $s \wedge s = s$ for any element s of S.
c) $s \vee t = t \vee s$ for any elements s and t of S.
d) If s, t, and w are any elements of S, then

$$(s \vee t) \vee w = s \vee (t \vee w).$$

Cite examples of some partially ordered sets we have touched on which are lattices. Cite examples which are not lattices. The set $P(S)$ of Example 9 is a lattice. Compare the above properties of $\vee$ and $\wedge$ in lattices with the propositions referring to $\cup$ and $\cap$ in Section 4.4. Why should we expect so much similarity?

chapter nine

More about Total Ordering

9.1 THE ORDERING OF THE INTEGERS, AN INTUITIVE VIEW

Chapter 8 introduced the term "a totally ordered set" to describe the integers, but the reader has been considering the integers as an ordered set since the earliest days of his schooling. In fact, a study of the integers is so much a part of elementary education that the reader must have formed many notions about what properties integers have and what properties they lack. The reader's intuitive notion of the integers has been the basis of many examples in this text. Although we now have a formal definition of the integers, and hence could prove much of what we formerly had to assume, we shall continue to view the integers more or less intuitively in order to prevent this text from becoming cumbersome.

The specific purpose of Section 9.1 will be to investigate the special order properties of integers. We begin with the fact that the set of integers is totally ordered by the relation defined by "is less than or equal to." What special properties does the total ordering of the integers have?

First, we note that there is no "first" integer, no integer less than all other integers, nor any "last" integer, no integer greater than all other integers. We note, however, that there is a first positive integer, namely 1, and a last negative integer, namely -1. (The integer 0 is not considered to be either positive or negative.)

Not all sets of positive real numbers have a first element as we see from the following.

Example 1. Even though there is a first positive integer, there is no first positive real number; that is, there is no positive real number less than all other positive real numbers. For if r is any positive real number, then $r/2$ is both positive and less than r; hence given any positive real number, we can always find a smaller positive real number.

Another property of integers is that every integer has both an immediate "successor" and an immediate "predecessor." That is, given any integer n, there is an integer that comes immediately after n (namely, $n + 1$), and an integer which comes immediately before n (namely, $n - 1$). There are no integers between n and $n + 1$, or between $n - 1$ and n. Such is not the case with the set of real numbers, as we see in the following example.

Example 2. Let r and r' be any real numbers with r strictly less than r'. Since $r < r'$ (the symbol $<$ denotes the phrase "is less than, but not equal to," that is, "strictly less than."), we have $r/2 < r'/2$. Adding $r/2$ to both sides of this inequality, we obtain

$$\frac{r}{2} + \frac{r}{2} = r < \left(\frac{r}{2} + \frac{r'}{2}\right).$$

On the other hand, if we add $r'/2$ to both sides, we obtain

$$\frac{r}{2} + \frac{r'}{2} < \left(\frac{r'}{2} + \frac{r'}{2}\right) = r'.$$

Thus $r/2 + r'/2$ is a real number which lies between r and r'; that is,

$$r < \left(\frac{r}{2} + \frac{r'}{2}\right) < r.$$

We note, too, that the immediate successor of any positive integer is also a positive integer, while the immediate predecessor of any negative integer is a negative integer. Moreover, the immediate predecessor of any positive integer except 1 (the first positive integer) is a positive integer, and the immediate successor of any negative integer except -1 (the last negative integer) is a negative integer.

Suppose n and m are any integers with n strictly less than m. The immediate successor of n is $n + 1$; the immediate successor of $n + 1$ is

$$(n + 1) + 1 = n + 2.$$

Starting at n, we can proceed to take immediate successors until we finally reach m. The important thing to note is that we *will* reach m, and, in fact, it will take us only finitely many steps ($m - n$ to be precise) to reach m. There are infinitely many integers, but between any two integers there are only finitely many integers. Thus, we can always proceed from one integer to any larger integer by taking finitely many immediate successors; we can also proceed from any integer to any smaller integer by taking finitely many immediate predecessors. However, the following example illustrates that even when we deal with a totally ordered set in which each element has an immediate successor, we cannot always get from any given element to any larger element in a finite number of steps.

Example 3. Let N be the set of positive integers and $M = \{0, 1\}$. Consider $M \times N$. Then $M \times N$ consists of all ordered pairs of the form (x, n) where x is either 0 or 1, and n is a positive integer. Typical elements of $M \times N$ would be $(0, 1)$, $(1, 5)$, $(0, 15)$, and $(1, 68)$. We define a relation R on $M \times N$ as follows: Any element of $M \times N$ with 0 as its first coordinate will be R-related to any element with 1 as its first coordinate. If two elements (x, m) and (x, n) of $M \times N$ have the same first coordinate, then we shall let (x, m) be R-related to (x, n) if m is less than or equal to n. It is left as an exercise to prove that R is a total ordering for $M \times N$. The elements of $M \times N$ ordered by R are represented as follows:

$$(0, 1), (0, 2), (0, 3), (0, 4), \ldots ; (1, 1), (1, 2), (1, 3), (1, 4), \ldots$$

Each element with first coordinate 0 precedes each element with first coordinate 1; elements having the same first coordinate are ordered according to the size of their second coordinate.

Note that each element $(0, n)$ has as immediate successor $(0, n + 1)$, and each element of the form $(1, n)$ has as immediate successor $(1, n + 1)$. Observe too that $(1, 1)$ has no immediate predecessor since the set of elements having 0 as first coordinate contains no last element. It is impossible to go from $(0, 1)$ to $(1, 1)$ in a finite number of steps by taking immediate successors, for starting at $(0, 1)$ and proceeding by immediate successors, we shall never reach the end of the elements that have 0 as first coordinate. Consequently, we have an example of a totally ordered set in which each element has an immediate successor, but such that we cannot always go from one element to any larger element in a finite number of steps by taking immediate successors.

Example 2 shows that no real number has an immediate successor; that is, given a real number r, there is no real number r' which immediately follows r (because there is always a real number between r and r'). Thus there is no possible way of proceeding from one real number to a larger real number by taking immediate successors.

There is one more significant property of the integers considered as a totally ordered set: In the set of integers, any nonempty subset W having a lower bound contains a first element. In other words, if W is a nonempty subset of the integers such that W has a lower bound, then W has a greatest lower bound v; moreover, v is an element of W. A little discussion about this property should cut through the rather forbidding language in which it must be formally stated.

Suppose W is a nonempty subset of the integers and has a lower bound. All we are saying is that W contains at least one integer and there is some integer k which is less than or equal to all the integers in W. If k is an element of W, then k must be the first element of W; hence k is the greatest lower bound of W. Suppose that k is not an element of W. Then starting at k and proceeding by immediate successors, $k + 1$, $k + 2$, $k + 3, \ldots$, we continue until we encounter an element of W; suppose $k + m$ is the first element of W that we reach. Then $k + m$ is an element of W which is less than or equal to every element of W, but less than any of the other elements of W; hence $k + m$ is a first element of W. Since $k + m$ is greater than or equal to every lower bound of W, $k + m$ is also the greatest lower bound of W. Consequently, W contains its greatest lower bound.

Since any nonempty set consisting solely of positive integers has 1 as a lower bound, we can say that any nonempty set of positive integers contains a least positive integer.

It is also true that any nonempty subset of integers which has an upper bound contains a last element (the last element being, in fact, the least upper bound), and that any nonempty subset of negative integers contains a greatest negative integer.

The following example shows that it is not generally true that any subset of a totally ordered set which has a greatest lower bound necessarily contains that greatest lower bound.

Example 4. Let P be the set of positive real numbers (that is, real numbers greater than 0) considered as a subset of the full set of real numbers. Then 0 is a lower bound for P, and 0 is not an element of P. Let r be any real number greater than 0. Then r is an element of P. Moreover, $r/2$ is an element of P which is even less than r, hence r could not be a lower bound of P. It follows then that any lower bound for P is less than or equal to 0. But this means that 0 is the greatest lower bound for P. The set P then has a greatest lower bound which is not an element of P.

Section 9.2 will be devoted to making more precise the notions we have handled rather informally in this section. The significance of the concepts we have been discussing thus far is that they lead to the *Principle of Finite Induction*. Therefore, we close this section with a statement of the principle and two informal proofs. We shall investigate the principle at some length and in slightly greater generality later in this chapter.

Proposition 1 (First Principle of Finite Induction). Suppose a property P holds true of the integer 1, and whenever an integer n has the property P, the integer $n + 1$ also has it. Then every positive integer has the property P.

Proof 1. Since 1 has the property P, and whenever n has the property P, $n + 1$ also has it, then $1 + 1 = 2$ has the property P. But then $2 + 1 = 3$ also has property P. Let m be any positive integer. Starting at 1 and proceeding by immediate successors, we reach m after only a finite number of steps; but at each step we obtain an integer which has the property P. Therefore m has the property P; consequently, each positive integer has the property P.

Proof 2. Let T be the set of positive integers for which the property P does not hold. Since 1 has the property P, 1 is not an element of T. Suppose T contains at least one positive integer. Then T contains a first positive integer m. Since m is the first element of T, $m - 1$ is a positive integer (since $m \neq 1$) and $m - 1$ is not an element of T. Therefore $m - 1$ has the property P. But if $m - 1$ has the property P, then

$$(m - 1) + 1 = m$$

also has the property P. This means, however, that m is an element of T which has the property P, a contradiction. Since the supposition that T contained at least one positive integer led to a contradiction, T must be empty. Consequently, the property P holds for every positive integer.

Example 5. To illustrate the use of Proposition 1, we prove:

If n is any positive integer, then $2 \cdot n$ is less than $3 \cdot n$.

Proof. Let P be the property that $2 \cdot n$ is less than $3 \cdot n$; then 1 has the property, since $2 \cdot 1 = 2$ is less than $3 \cdot 1 = 3$. Now we show that if a positive integer n has the property P, then $n + 1$ also has it. Suppose n has the property P, that is, suppose $2 \cdot n$ is less than $3 \cdot n$. Adding 2 to both sides of this inequality, we have

$$2 \cdot n + 2 = 2 \cdot (n + 1) \text{ is less than } 3 \cdot n + 2.$$

But

$$3 \cdot n + 2 \text{ is less than } (3 \cdot n + 2) + 1 = 3 \cdot n + 3 = 3 \cdot (n + 1).$$

Hence $2 \cdot (n + 1)$ is less than $3 \cdot (n + 1)$. This is the property P for $n + 1$; hence whenever property P holds for n, it also holds for $n + 1$. By Proposition 1, then, property P is true for each positive integer.

EXERCISES 9.1

1. Prove that the relation R in Example 3 totally orders $M \times N$.
2. Prove that the totally ordered set $M \times N$ in Example 3 has the property that every nonempty subset contains a first element. [*Hint:* Let W be a nonempty subset of $M \times N$. Prove that if

$$W \cap \{(0, n) \mid n \text{ a positive integer}\}$$

is nonempty, then this intersection contains a first element which is the first

element of W. If
$$W \cap \{(0, n) \mid n \text{ a positive integer}\} = \varnothing,$$
then prove that
$$W \cap \{(1, n) \mid n \text{ a positive integer}\}$$
contains a first element which is the first element of W. Use the fact that with regard to order, the sets
$$\{(0, n) \mid n \text{ a positive integer}\}$$
and
$$\{(1, n) \mid n \text{ a positive integer}\}$$
are essentially the same as the set of positive integers.

3. Using the formal definition of integer given in Section 8.3 and the total ordering of the integers described in Exercise 6, Section 8.3, prove each of the following:

   a) There is no integer $|\overline{m}, \overline{n}|$ such that $|\overline{\varnothing}, \overline{\varnothing}| < |\overline{m}, \overline{n}| < |\overline{1}, \overline{\varnothing}|$. [*Hint:* You will use the fact that there is no set "between" $\varnothing$ and a set consisting of exactly one element.]

   b) Prove that if $|\overline{m}, \overline{n}|$ is any integer, then $|\overline{m}, \overline{n}| < |\overline{m} + \overline{1}, \overline{n}|$. Moreover, there is no integer $|\overline{p}, \overline{q}|$ such that $|\overline{p}, \overline{q}|$ lies "between" $|\overline{m}, \overline{n}|$ and $|\overline{m} + \overline{1}, \overline{n}|$; that is, prove that $|\overline{m} + \overline{1}, \overline{n}|$ is the immediate successor of $|\overline{m}, \overline{n}|$. What is the immediate predecessor of $|\overline{m}, \overline{n}|$?

4. Part (a) of Exercise 3 says essentially that there is no integer between 0 and 1; that is, there is no integer n such that $0 < n < 1$. Prove that if there is an integer n such that $0 < n < 1$, then there is no least integer greater than 0. Prove that if there is no integer n such that $0 < n < 1$, then given any positive integer m, there is no integer between m and $m + 1$.

5. Using the Principle of Finite Induction, prove each of the following:

   a) If n is any positive integer, then $n < n + 1$.
   b) If n is any positive integer, then $n^2 + n$ is evenly divisible by 2.

   In (a) and (b), we must first prove the given statement for $n = 1$. Then we must show that if the statement is true for n, it must also be true for $n + 1$. (Note that we *assume* the truth of the statement for n and must prove that this assumption leads to the truth of the statement for $n + 1$.) There will be more about finite induction later in this chapter.

6. Produce an argument to show that every nonempty subset of integers which has an upper bound contains a last element. Prove that if a property P is true for -1, and if the property's being true for any integer n implies that the property is also true for $n - 1$, then the property is true for every negative integer.

7. Prove that every nonempty subset of the integers which has both an upper bound and a lower bound contains finitely many elements. Is this property true for the real numbers; that is, is it true that any subset of real numbers is finite if it has an upper bound and a lower bound?

9.2 THE ORDERING OF THE INTEGERS, A FORMAL VIEW

Let M be the set of even integers, that is,

$$M = \{2k \mid k \text{ is an integer}\}.$$

Letting M be totally ordered by the relation defined by "is less than or equal to," we can show that M has virtually the same order properties as the full set of integers. For example, each element $2k$ of M has an immediate successor $2k + 2$ and an immediate predecessor $2k - 2$. Starting at any element of M and taking immediate successors, we reach any larger element of M in only a finite number of steps. In other words, M is essentially the same totally ordered set as is the set of integers. The following discussion attempts to define this type of totally ordered set more precisely.

Conventions to be observed in the sequel. Unless otherwise indicated, a set S is one that is totally ordered by a relation R. We assume for the sake of simplicity that R is generated by the phrase "is less than or equal to." If s and t are any elements of a set S totally ordered by R, then *s is less than t* will mean that s is less than or equal to t, but not equal to t. We shall use *s is greater than t* to indicate that t is less than s. We also assume that any set of numbers, for example, the integers, has the "less than or equal to" ordering.

Definition 9.1. *(Let S be a set totally ordered by R.) If s is any element of S, then an element t of S is said to be a successor of s if s is less than t. The element t is said to be an immediate successor of s if (1) t is a successor of s, and (2) there is no element w of S such that w is greater than s but less than t; that is, there is no element of S between s and t.*

An element t of S is said to be a predecessor of the element s of S if t is less than s. We say that t is an immediate predecessor of s if (1) t is a predecessor of s, and (2) there is no element w of S such that w is greater than t but less than s.

In other words, t is the immediate successor of s if there is no successor of s less than t; and t is the immediate predecessor of s if there is no predecessor of s greater than t. We will denote the immediate successor (if one exists) of the element s by s^+, and the immediate predecessor of s (if one exists) by s^-. We now prove certain basic properties about immediate successors and predecessors.

Proposition 2. *If t and t' are both immediate successors of s, then $t = t'$. (That is, if an element s has an immediate successor t, then t is the only immediate successor of s.)*

Proof. Since S is a totally ordered set (see the paragraph of conventions above), if $t \neq t'$, then either t is less than t', or t' is less than t. If t is less than t', then t is a successor of s which is less than t', a fact that contradicts the assumption that t' is an immediate successor of s. On the other hand, if t' is less than t, then t' is a successor of s which is less than t, contradicting the assumption that t is an immediate successor of s. Since both "t less than t'" and "t' less than t" lead to contradictions, it must be that $t = t'$.

Proposition 3. $(s^+)^- = s$; *that is, the immediate predecessor of the immediate successor of s is s.*

Proof. Since s is less than s^+, s is a predecessor of s^+. If s is not the immediate predecessor of s^+, then there is an element t such that s is less than t, but t is less than s^+. However, this means that t is a successor of s which is less than s^+, the immediate successor of s—an impossibility. Therefore s must be the immediate predecessor of s^+; that is,

$$(s^+)^- = s.$$

We formalize the order properties of the integers in the following definition.

Definition 9.2. *A totally ordered set S is said to be a discrete set if it has the following properties:*

D1. *Every element of S which has any predecessor has an immediate predecessor.*

D2. *Any element of S which has any successor has an immediate successor.*

D3. *If s and t are any elements of S such that s is less than t, then by taking s^+, $(s^+)^+$, $((s^+)^+)^+$, etc., we reach t in a finite number of steps. That is, starting at any element s of S and taking immediate successors, we can reach any element of S greater than s in a finite number of steps.*

An element of S which has no predecessors is called a first element; any element of S which has no successors is called a last element.

A discrete set with a first element but no last element is called a progression; a discrete set with a last element but no first element is called a regression.

Example 6. The set of negative integers (totally ordered by the relation defined by "less than or equal to") forms a regression with last element -1. The set of positive integers forms a progression with first element 1. The full set of integers forms a discrete set with neither first nor last element.

Example 7. The set of positive even integers (with the "less than or equal to" total ordering) also forms a progression, the first element being 2. The full set of even integers forms a discrete set without first or last elements.

Suppose a totally ordered set S is a progression. Then S has a first element which we denote by 1. If 1 has any successor, it has an immediate successor 1^+ (by Property D2); we will denote 1^+ by 2. If 1 has no successor at all, then $S = \{1\}$. If 2^+ exists, we may denote it by 3; if 2^+ does not exist, then $S = \{1, 2\}$. We can continue in like manner, denoting $(n-1)^+$ by n if $n-1$ has a successor; otherwise,

$$S = \{1, 2, 3, \ldots, n-1\}.$$

The point is that any progression essentially looks like a subset of the positive integers (totally ordered by the "less than or equal to" relation); if S has no last element, then, with regard to order, we can make S look exactly like the set of positive integers by suitably choosing our notation for the elements of S. Every element of S will be labeled with the symbol for some integer since each element of S is reached after only finitely many steps.

There are important progressions without any last element which we do not wish to confuse with the set of positive integers, thus we shall denote the first term of an arbitrary progression by s_1, s_1^+ by s_2, s_2^+ by s_3; in general, the "nth" term of a progression S will be denoted by s_n.

We should also observe that the elements of any discrete set S which has no first or last elements can be "labeled" so as to make S look like the set of integers. Specifically, choose any element s of S and denote s by 0. Denote s^+ by 1 and s^- by -1. Denote 1^+ by 2 and $(-1)^-$ by -2, etc. Thus, S becomes

$$\ldots, -3, -2, -1, 0, 1, 2, 3, 4, \ldots$$

Each element of S is assigned a symbol since each element is reached after only finitely many steps.

If, in fact, any discrete set without first or last elements can be made to look like the integers, we might wonder whether the integers might be defined as any discrete set without first or last elements. From the point of view of order alone, the full set of integers (with the "less than or equal to" total ordering) can be characterized in this manner. That is, the order properties of the integers are essentially found in any discrete set without first or last elements. However, there is more to the integers than just the order properties. For example, we can add and multiply any two integers. With regard to order, the set of even integers has the same properties as the full set of integers, but concerning addition and multiplication, these sets are different. For example, there is no even integer k such that for any even integer n, $k \cdot n = n$. There is, however, an integer 1 with the property that $1 \cdot n = n$ for any integer n.

Suppose we begin, however, with a discrete set S without first or last elements and ask the following question: Can we by means of the *order properties alone* of S define an addition and multiplication for S so that even from an algebraic point of view, S looks like the integers? We will answer this question after a further study of finite induction.

EXERCISES 9.2

1. Prove that if t and t' are both immediate predecessors of s, then $t = t'$ (see Proposition 2).

2. Recall that we are working in the context of a totally ordered set S. Prove each of the following:
   a) $(s^-)^+ = s$, for any element s of S which has an immediate predecessor.
   b) S contains at most one first element.
   c) S contains at most one last element.

3. Prove that any discrete set which has both a first and a last element contains only finitely many elements. Prove that any discrete set which contains only finitely many elements must contain both a first and a last element.

4. Each of the following sets of numbers is to be considered ordered by the "less than or equal to" relation. Which of these sets is a discrete set? a progression? a regression? If a set is not discrete, indicate which of the properties D1, D2, and D3 of Definition 9.2 the set fails to satisfy.
   a) $\{1, 2\}$
   b) $\{1, 1.1, 1.11, 1.111, 1.1111, 1.11111, \ldots\}$
   c) $\{n \mid n \text{ is an integer less than } -6\}$
   d) $\{q \mid q \text{ is the quotient of two positive integers}\}$ For example, typical elements of this set would be $\frac{1}{2}$, $\frac{5}{6}$, $\frac{9}{17}$, and $\frac{67}{189}$.
   e) $\{1/n \mid n \text{ is a nonzero integer}\}$
   f) $\{1 - 1/n \mid n \text{ is a positive integer}\} \cup \{2 - 1/n \mid n \text{ is a positive integer}\}$
   g) $\{d \mid d \text{ is an unending decimal whose digits are either 0 or 1}\}$

5. The pages of a book form a discrete set that one encounters daily. Find other examples of discrete sets that might be encountered in everyday living.

6. Prove that any discrete set is countable.

9.3 FINITE INDUCTION*

In Section 9.1, Proposition 1, we presented the First Principle of Finite Induction. This principle is a statement that yields a method of proving

---

* The paragraph on *Conventions* of Section 9.2 holds true for the rest of Chapter 9.

properties of the positive integers, and its proof depends solely on the order properties of the positive integers. Since a progression is a totally ordered set having essentially the same order properties as those of the set of positive integers (with the sole exception that a progression might have a last element, while the positive integers, of course, do not), we should expect that a comparable principle must exist for progressions. Such is indeed the case; the statement and proofs for this principle are essentially the same as those for the First Principle of Finite Induction for integers.

Proposition 4. *Let S be any progression; denote the first element of S by s_1. Suppose s_1 has some property P, and whenever an element s of S has property P, s^+, the immediate successor of s, also has property P. Then every element of S has property P.*

(The reader should carefully compare this proposition with Proposition 1, which dealt with one specific progression, the set of positive integers; Proposition 4 deals with progressions in general. The proofs of Proposition 1 were based on intuitive ideas about the order properties of the set of integers, while Proposition 4 can be proved rigorously from the formal definition of a progression.)

Proof 1. Since s_1 has the property P (and whenever an element s of S has the property P, s^+ also has the property P), s_1^+ has the property P. But then $(s_1^+)^+$ also has the property P. Let s' be any element of S. Starting at s_1 and proceeding by immediate successors, we reach s' after only a finite number of steps; but at each step we obtain an element of S which has the property P. Therefore s' has the property P; consequently, each element of S has the property P.

We shall look at a second proof for Proposition 4 after proving the following proposition.

Proposition 5. *Let S be a progression and T be any nonempty subset of S. Then T contains a first element; that is, T contains an element which is less than any other element of T.*

Proof. If s_1, the first element of S, is an element of T, then s_1 is less than any other element of T because s_1 is less than any other element of S. If s_1 is not an element of T, then proceed to $s_1^+ = s_2$. If s_2 is an element of T, then s_2 is less than any other element of T because s_2 is less than any other element of S except s_1, but s_1 is not an element of T. If s_2 is not an element of T, then we proceed by immediate successors until we reach s_n, the least element of S which is an element of T. Then s_n is less than any other element of T, because s_n is less than any other element of

S except $s_1, s_2, \ldots, s_{n-1}$, but these are not elements of T. Since T contains at least one element, we are certain to encounter such an s_n in only a finite number of steps.

A second proof for Proposition 4. Let T be the set of elements of S for which the property P does not hold. Since s_1, the first element of S, has the property P, s_1 is not an element of T. Suppose T is nonempty; that is, at least one element of S does not have the property P. Then by Proposition 5, T contains a first element t. Since s_1 is a predecessor of t, t has an immediate predecessor t^- (note that we could not be sure that t had a predecessor if it were possible for s_1 to be an element of T). Since t is the first element of T, t^- is not an element of T; hence t^- has the property P. But if t^- has the property P, then $(t^-)^+ = t$ also has the property P; consequently, t is not an element of T, a contradiction. Since the supposition that T was nonempty led to a contradiction, T must be empty. Consequently, each element of S has the property P.

Example 8. Let $S = \{s_1, s_2, s_3, \ldots\}$ be the progression formed as follows: We define s_1 to be the sum of the first one positive integer, that is, $s_1 = 1$. We define s_2 to be the sum of the first two positive integers, that is, $s_2 = 1 + 2 = 3$. In general, s_n, the nth term of S, will be the sum of the first n positive integers, $1 + 2 + 3 + \cdots + (n - 1) + n$. We note that

$$s_1 = 1 = \frac{1 \cdot (1 + 1)}{2}$$

and

$$s_2 = 1 + 2 = 3 = \frac{2 \cdot (2 + 1)}{2}$$

and

$$s_3 = 6 = \frac{3 \cdot (3 + 1)}{2}.$$

Using Proposition 4, we shall prove we can find the nth term of S, that is, the sum of the first n positive integers, by the following formula:

1. $s_n = \dfrac{n \cdot (n + 1)}{2}.$

We saw that formula (1) is valid for the first element of S, and it is also valid for the immediate successor of the first element of S, and for the next immediate successor. We now show that if the formula is valid for the element s_n of S, it will also be valid for

$$s_n^+ = s_{n+1}.$$

Let us suppose then that formula (1) is valid for s_n, that is,

$$s_n = \frac{n \cdot (n + 1)}{2}.$$

Now we define s_{n+1} as

$$s_{n+1} = (\text{df}) \ 1 + 2 + 3 + \cdots + (n-1) + n + (n+1) = s_n + (n+1).$$

That is, s_{n+1} is the sum of the first n positive integers plus $n+1$. Since $s_{n+1} = s_n + (n+1)$ and since formula (1) is valid for s_n, we have

2. $\quad s_{n+1} = s_n + (n+1) = \dfrac{n \cdot (n+1)}{2} + (n+1)$

$$= \frac{n \cdot (n+1) + 2 \cdot (n+1)}{2}$$

$$= \frac{(n+2) \cdot (n+1)}{2}$$

$$= \frac{(n+1) \cdot ((n+1)+1)}{2}.$$

The last term in (2), however, is precisely formula (1) with $n+1$ substituted for n. We have thus shown that if formula (1) is valid for s_n, it will also be valid for s_{n+1}. Since we know that formula (1) is valid for s_1, we can say that by Proposition 4, formula (1) is valid for every element of S. We have thus proved: The sum s_n of the first n positive integers is given by the formula

$$s_n = \frac{n \cdot (n+1)}{2}.$$

Example 9. We shall prove: If n is any integer greater than 3, then $2 \cdot n + 1$ is less than $3 \cdot n - 1$.

Proof. To say that n is an integer greater than 3 is the same as saying that n is an element of the progression $S = \{4, 5, 6, 7, 8, \ldots\}$. The property we wish to prove of each element of this progression is

3. $\quad 2 \cdot n + 1$ is less than $3 \cdot n - 1$.

Since $2 \cdot 4 + 1 = 9$ is less than $3 \cdot 4 - 1 = 11$, property (3) holds true for the first element 4 of S. We now prove that if the property (3) applies to the element n of S, it will also apply to $n+1$.

Let us assume that property (3) holds for the element n of S; that is, we assume that $2 \cdot n + 1$ is less than $3 \cdot n - 1$. We argue as follows:

4. $\quad 2 \cdot n + 1$ is less than $3 \cdot n - 1$. (*Assumed.*)

5. $\quad 2 \cdot n + 1 + 2 = 2 \cdot (n+1) + 1$ is less than

$$3 \cdot n - 1 + 2 = 3 \cdot n + 1.$$

[*Adding 2 to both sides of the inequality (4).*]

6. But $3 \cdot n + 1$ is less than

$$(3 \cdot n + 1) + 1 = 3 \cdot n + 2 = 3 \cdot n + 3 - 1 = 3 \cdot (n + 1) - 1.$$

7. Therefore $2 \cdot (n + 1) + 1$ is less than $3 \cdot (n + 1) - 1$. [*Shown by (5) and (6); (7) is precisely property (3) applied to $n + 1$.*]

By Proposition 4, therefore, property (3) holds for any element of S, that is, for any integer greater than 3.

In any proof using the generalized First Principle of Finite Induction (Proposition 4) two things must be shown: (1) that the property in question holds for the first element of the progression, and (2) if the property holds for some element of the progression, it must hold for the immediate successor of that element. If we cannot show either (1) or (2), we do not have a valid proof. We can see why we need both criteria satisfied when we look at the following examples.

Example 10. Let S be the progression of integers greater than -9, that is, $S = \{-8, -7, -6, -5, \ldots\}$. It is true that if an element n of S is positive, then $n + 1$ must be positive. If it were also true that the first element -9 of S was positive, then according to Proposition 4, every element of S would be positive—which, of course, is not the case. If we tried to prove by induction that each element of S was positive we would fail because the first element of S is not positive.

Example 11. Let S be the set of positive integers. Then the first element 1 of S is odd. This does not mean, however, that every element of S is odd because it is not true that if n is an odd element of S, then $n + 1$, the immediate successor of n, is also odd.

EXERCISES 9.3

1. Prove each of the following by means of finite induction.
   a) If n is any positive integer, then n is less than $2 \cdot n$.
   b) If n is any positive integer, then $4 \cdot n$ is greater than $2 \cdot n + 1$.
   c) The sum of the first n positive even integers is given by $n \cdot (n + 1)$. [*Hint:* The nth positive even integer is $2 \cdot n$, thus the sum of the first n positive even integers is $2 + 4 + 6 + \cdots + 2 \cdot (n - 1) + 2 \cdot n$.]
   d) The sum of the first n positive odd integers is n^2. [*Hint:* The nth positive odd integer is $2 \cdot n - 1$. For example, the first positive odd integer is $2 \cdot 1 - 1 = 1$, and the second is $2 \cdot 2 - 1 = 3$. Thus the sum of the first n positive odd integers is $1 + 3 + 5 + \cdots + (2 \cdot n - 1)$.]
   e) If n is any positive integer, then n is less than $n^2 + 1$.

f) The sum of the first n perfect squares, that is, $1^2 + 2^2 + 3^2 + \cdots + n^2$, is $\frac{1}{6} \cdot n \cdot (n+1) \cdot (2 \cdot n + 1)$.

g) If n is any positive integer larger than 4, then n^2 is less than 2^n. [*Note:* It follows that if S is a set of n elements, then there are more subsets of S than elements of $S \times S$.]

h) If n is an integer greater than 2, then 3 evenly divides $n^3 - n$.

2. The *Second Principle of Finite Induction* may be stated as follows: Let S be any progression, and let s_1 be the first element of S. Then if some property P is true of s_1 and whenever the property P is true of all predecessors of any element s of S, it is true also for s, then the property P is in fact true for any element of S. Supply a proof.

3. Use the Second Principle of Finite Induction to prove the following: If n is any positive integer greater than or equal to 2, then the sum of n positive integers is again a positive integer. [*Hint:* If $a_1, a_2, \ldots, a_n$ are any n positive integers, then $a_1 + a_2 + \cdots + a_n = (a_1 + a_2) + (a_3 + a_4 + \cdots + a_n)$.]

4. State and prove a Principle of Finite Induction for regressions (see Definition 9.2). [*Hint:* You might begin "If a property P is true for the *last* element of a regression S and whenever the property P is true for an element s of S, it is true also for $s^-, \ldots$"]

5. Prove each of the following:

a) The product of n positive integers is again a positive integer if n is greater than or equal to 2.

b) If a and b are two real numbers and m is any positive integer, then

$$(a \cdot b)^m = a^m \cdot b^m.$$

9.4 ANOTHER DEFINITION OF THE INTEGERS

Let S be any discrete totally ordered set without first or last elements. We saw in Section 9.2 that by properly "labeling" the elements of S, we could make S look like the set of integers (at least with regard to notation and order). That is, we have S represented as

$$\ldots, -4, -3, -2, -1, 0, 1, 2, 3, 4, \ldots$$

We now seek to define addition and multiplication on S in such a way that S resembles the integers with regard not only to order and notation, but to its "arithmetic" structure as well.

Given any two elements s and t of S, we have labeled the elements of S in such a way that s and t have each been assigned the *symbol* for some integer. For example, suppose s has been labeled 5 and t has been labeled 8. Then we might "define" $s + t$ to be that element of S which we have labeled with $5 + 8 = 13$. Similarly, we might define $s \cdot t$ to be that element of S which has been labeled by $5 \cdot 8 = 40$. With addition and multiplication so defined and with the notation we have adopted, it would be quite impossible to distinguish S from the actual and true set of integers. But if we defined addition and multiplication in this way, we would really be cheating. Why? Because this method of defining the sum and product of two elements of S is really only choosing what we already know to be arithmetically appropriate labels for the elements of S. We would be begging a very important question, namely, what are the integers?

We have noted that a discrete totally ordered set without first or last elements is very nearly the same as the set of integers, or, at least, of our intuitive notion of the integers. Actually, however, our knowing what a discrete totally ordered set is does not in any way depend on our knowing what the integers are. If we now can define addition and multiplication on any discrete totally ordered set in such a way that these operations are essentially the same as addition and multiplication in the integers, and if we can define addition and multiplication *without* appealing to what we already know about the integers, then we shall be very close indeed to actually having a genuine definition of the integers. That is, we could then define the integers to be any discrete set without first or last elements and with the operations of addition and multiplication as we shall soon define them.

Defining $s + t$ to be that element of S which is labeled by $5 + 8 = 13$ presumes that we already know that $5 + 8$ should be 13; that is, it presumes that we already know precisely what the rules for addition and multiplication of integers are; and these rules, in turn, lead to the following circular definition of the integers: The integers are any discrete totally ordered set which does not have first or last elements and for which addition and multiplication are defined as for the integers. This does not mean that when we finish defining addition and multiplication for a discrete set we will not find that $5 + 8 = 13$. But the fact that $5 + 8 = 13$ will be a consequence of the definition of addition and not the definition itself. Addition and multiplication will be defined entirely in terms of the *order* properties of a discrete set.

Before giving another definition of the integers, we might ask what relationship there is between this definition based on the properties of a certain type of totally ordered set and the definition of the integers, Definition 8.4, which was based on the order and arithmetic properties of cardinal numbers. Definition 8.4 was, however, merely another way of looking at the same thing. The properties of the integers can be proved either from Definition 8.4 or from the definition to be presented later in this section; the choice of the definition to use in any study of the integers is, to some degree, a matter of taste. Imagine, for example, an enormous mountain. Perhaps one surveying party views the mountain from the south, another from the west, and still another party flies over it. The mountain remains the same, but the viewpoint of each surveying party is a little different. The integers remain the same, but our approach to them can vary.

This once again raises the interesting and extremely profound question: Are mathematical notions really *a priori*, existing prior to and apart from our efforts to define them, or do mathematical objects come into being only when a mathematician embodies them in a suitable definition? We are no more prepared to explore the intricate depths of this question now than when we first mentioned it in Section 8.3.

We now proceed to define addition and multiplication for a discrete set S without first or last elements. In order to avoid giving the impression of circularity in our definition, we will not label the elements of S, using symbols for integers alone. Instead, we shall choose one element of S and denote it by s_0. We will denote s_0^+ by s_1, and s_0^- by s_{-1}. We denote s_1^+ by s_2, and s_{-1}^- by s_{-2}. In other words, we proceed in exactly the same manner as when we labeled the elements of S to make S look exactly like the set of integers, except that in this instance the integers appear as subscripts. Thus S will look like

$$\ldots, s_{-5}, s_{-4}, s_{-3}, s_{-2}, s_{-1}, s_0, s_1, s_2, s_3, s_4, s_5, s_6, \ldots$$

The relabeling has been done to emphasize the fact that we are not presupposing any prior knowledge of the integers. In fact, we could define addition and multiplication without appealing to any symbols for any integers whatever. Although avoiding all use of any symbols for integers would emphasize even more strongly that our definition is not circular, it would also make matters so complex notationally that the technique is not feasible for this text.

We define addition of elements of S by the following:

A1. *For any element s of S,*

$$s + s_0 = (\text{df}) \; s_0 + s = (\text{df}) \; s.$$

A2. *If s_k is any element of S, then*

$$s_k + s_1 = (\text{df}) \; s_1 + s_k = s_k^+ = (\text{df}) \; s_{k+1}.$$

We let

$$s_k + s_2 = (\text{df}) \; s_2 + s_k = (\text{df}) \; s_k + (s_1 + s_1)$$
$$= (s_k + s_1) + s_1 = (\text{df}) \; s_{k+1} + s_1 = s_{k+1}^+ = s_{k+2}.$$

In general, for any element s_k of S, and any element s_n of S greater than s_0, we define

$$s_k + s_n = (\text{df}) \; (s_k + s_{n-1})^+ = (\text{df}) \; s_{k+n}.$$

Note that we now can add any element of S to any element of S which is greater than or equal to s_0; this is possible because we are told in A1 how to add s_0 to any element of S, and in A2 how to add s_1 to any element of S. We are also told in A2 how to add s_n to any element s_k of S, given $s_k + s_n^-$, where s_n is greater than s_0. Since we know how to add s_0 and s_1, we know how to add s_2, hence we know how to add s_3, etc. Thus, we now know how to add any two elements of S provided that at least one of them is greater than or equal to s_0. In A3 we shall find out how to add the "negative" elements of S.

A3. *For any element s of S,* $s + s_{-1} = (\mathrm{df})\ s^- = (\mathrm{df})\ s_{-1} + s$. *For any element s_k of S and any element s_{-n} of S less than s_0, we set*

$$s_k + s_{-n} = (\mathrm{df})\ s_{-n} + s_k = (\mathrm{df})\ (s_{-n+1} + s_k)^- = s_{-n+k}.$$

The first sentence of A3 tells us how to add s_{-1} to any member of S. Then it gives a rule such that knowing $s_k + s_{-1}$, we can determine

$$s_k + s_{-1}^- = s_k + s_{-2}.$$

Knowing this sum, we can then determine $s_k + s_{-3}$, and hence $s_k + s_{-4}$. Thus we can determine the sum of any element of S with any element of S less than s_0.

Together, (A1), (A2), and (A3) tell us how to add any two elements of S; moreover these rules were stated only in terms of the ordering of S and did not depend on knowing what the sum of two integers is.

We now define the product of two elements of S.

M1. *For any element s of S,*

$$s \cdot s_0 = (\mathrm{df})\ s_0 \cdot s = (\mathrm{df})\ s_0.$$

M2. *For any element s of S,*

$$s_1 \cdot s = (\mathrm{df})\ s \cdot s_1 = (\mathrm{df})\ s \qquad \text{and} \qquad s_2 \cdot s = (\mathrm{df})\ s \cdot s_2 = (\mathrm{df})\ (s_1 \cdot s) + s$$

(remember that addition has already been defined). In general, if s is any element of S and s_n is any element of S greater than s_0, we set

$$s_n \cdot s = (\mathrm{df})\ s \cdot s_n = (\mathrm{df})\ (s \cdot s_{n-1}) + s.$$

Since multiplication is defined for s_1, it is defined for s_2, and hence for s_3, etc.; thus multiplication is defined for any element of S greater than s_0.

M3. *For any element s_k of S, we set*

$$s_k \cdot s_{-1} = (\mathrm{df}) \ s_{-1} \cdot s_k = (\mathrm{df}) \ s_{-k}.$$

As a rule, for any element s_k of S, and for any element s_{-n} of S less than s_0, we set

$$s_k \cdot s_{-n} = (\mathrm{df}) \ s_{-n} \cdot s_k = (\mathrm{df}) \ s_{-n+1} \cdot s_k + s_{-k}.$$

Note that addition and multiplication have both been *defined* so that s_1 acts like 1, s_0 acts like 0, and

$$s + t = t + s \text{ and } s \cdot t = t \cdot s$$

for any elements s and t of S. We thus arrive at the following definition of the integers.

Definition 9.3. *A set Z which is discrete (and totally ordered) without first or last elements, and with addition and multiplication as defined above is said to be the set of integers.*

Definition 9.3, like Definition 8.4 can be used as the basis for a formal study of the integers.

EXERCISES 9.4

1. In order that (A1), (A2), and (A3) define the sum of any two elements of S, the following must be true: If the first element of a progression (last element of a regression) is given together with a rule that defines any element of the progression (regression) in terms of its predecessors (successors), then each element of the progression (regression) is uniquely determined. For example, suppose the first element of a progression S is 3. Suppose we are told that any element s of S (except the first element) is equal to $s^- + 1$. Then s_2, the second element of S, is computed to be

$$s_2 = s_2^- + 2 = s_1 + 2 = 3 + 2 = 5.$$

Similarly, $s_3 = s_2 + 2 = 5 + 2 = 7$. In the following we give the first element of a progression together with a rule for determining any element when its predecessors are given. Find the first five terms in each progression.

a) $s_1 = 2, \ s_n = s_{n-1} + 1$
b) $s_1 = 0, \ s_n = s_{n-1} + 4$
c) $s_1 = 2, \ s_n = (s_{n-1})^2$

2. Prove that if s, t, and w are integers, then

a) $s + t = t + s$ b) $(s + t) + w = s + (t + w)$.

chapter ten

Probability

10.1 THE NATURE OF PROBABILITY

If we take a fair penny and flip it fairly, we have a 50–50 chance of having the penny come up heads. The argument to this conclusion is more philosophical than mathematical. When flipping a coin, we can obtain either heads or tails on any given toss. However, if both the coin and the flip are fair, that is, not weighted in favor of either heads or tails, we have no reason for supposing that heads will appear with greater frequency than tails. We are thus presented with two possible outcomes of any toss, neither of which is in any way more probable than the other. Consequently, we must assign the same chance of occurrence to each (since we have no reason to assign a greater chance of occurrence to one of them). We conclude then that heads should occur half the time, and tails should occur half the time; that is, heads and tails each have a 50–50 chance of occurring.

Suppose we have a fair die (with faces 1 through 6). If we roll the die a number of times, we might reasonably expect that the face 1 will occur about one roll in six; for example, if we roll the die 36 times, we might expect about six 1's. The reason for this is that there are six faces to the die and we have no reason to suppose that any one face will appear more often than another. Since there are six faces, each face should then occur on about $\frac{1}{6}$ of the rolls.

So that we can be more precise, we make the following definitions.

Definition 10.1. An experiment is any clearly defined procedure to be carried out; that is, an experiment is some precise action that can be performed. A trial is an actual carrying-out, or performance, of an experiment. An event is a possible outcome, or result, of a trial.

The following examples illustrate the distinction between these terms.

Example 1. If we have some particular coin, one experiment would be to raise the coin to eye level, drop it on the floor, and let it come to rest unhindered. A trial would be an actual performance of this procedure. Each time we repeat the experiment we would be performing the same experiment, but a different trial. The particular events in this case would be "the coin comes up heads" or "the coin comes up tails." Were it truly possible for the coin to remain on its side, then this would also have to be taken into account as an event.

Example 2. Taking some particular die and rolling it on a table would be an experiment. A particular trial in this case would be one actual roll of the die. The possible outcomes, or events, would be each of the six faces of the die.

For a trial that has several possible outcomes, probability theory is interested in finding some quantitative measure which will tell us what chance any given event has of occurring. Since we expect heads to occur half the time if we are flipping a fair coin, we can reasonably assign probability $\frac{1}{2}$ to the event "heads." Likewise, $\frac{1}{6}$ seems an appropriate measure of the chance that a 1 will occur in any roll of a fair die.

Even if the probability of heads is $\frac{1}{2}$, this does not necessarily mean that heads will appear in exactly $\frac{1}{2}$ of the trials performed. For example, if we flip a fair coin fairly 10 times, it is really not very likely that we will get precisely 5 heads and 5 tails; it is even possible that we will get 10 heads or no heads at all. What we *can* say though is that if our assignment of $\frac{1}{2}$ to the event "heads" is accurate, then in the long run, that is, as we accumulate the results of more and more trials, heads and tails should occur proportionately more and more equally.*

Working in the other direction, if we did not know what the probability of some particular event should be, but we were able to perform an appropriate experiment a great many times, then we might be able to arrive at an estimate of the probability of the event. Suppose we perform the experiment 100 times and find that the event in question occurs 35

---

* We say "proportionately more and more equally" to indicate that the ratio of heads to the total number of flips approaches $\frac{1}{2}$, even though we may have a large difference in the number of heads vs. tails. For example, if we flipped the coin 10 times and got 4 heads and 6 tails, the ratio of heads to the total number of flips is $\frac{4}{10}$, but the difference between the number of heads and tails is 2. If we flip 1000 times and get 475 heads and 525 tails, then the ratio of heads to total flips is $\frac{475}{1000}$ (which is much closer to $\frac{1}{2}$ than $\frac{4}{10}$ is) even though we now have 50 more tails than heads.

times. We can then say that the event occurs about 35% of the time, that is, has a 35% chance of occurring; hence 0.35 is a reasonable figure to assign as the probability of the given event. Since 100 trials represent actually a fairly small number of trials, our first estimate might be well off the true probability. Suppose we repeat the experiment another 900 times and find that the event has occurred 310 times in 1000 trials. We might then feel justified in saying that 0.310 is a better estimate than 0.35 of the true probability of the event. This procedure of computing probabilities by experiment is actually quite common. For example, if a weatherman says there is an 80% chance of rain, what he is saying is that in 80% of the cases in the past in which a weather pattern similar to the one in question occurred, rain also occurred. The particular weather pattern is the trial and rain is the event. The probability of rain is estimated to be the ratio of times rain has occurred to the total number of times the particular weather pattern has occurred.

If we are not sure what the probability of some event E is, but we are able to perform an appropriate experiment, then

1. The probability of $E \approx \dfrac{\text{number of trials in which } E \text{ occurs}}{\text{total number of trials performed}}$,

and the more trials that are performed, the better the approximation is likely to be.

Definition 10.2. *Suppose E is any event, and we have performed n trials; in m of them, E has occurred. The ratio m/n is called the relative frequency of E, denoted by $R(E)$.*

As n becomes larger and larger, the relative frequency of E should get closer and closer to the probability of E.

Example 3. Suppose a certain coin is flipped 1000 times and 546 heads appear. Then the relative frequency of heads (relative to this particular set of 1000 trials) is $546/1000 = 0.546$. Suppose that the coin is flipped 1,000,000 times and 502,000 heads occur; then R(heads) is now 0.502.

Definition 10.2 implicitly states that m is an integer between 0 (if E does not occur at all) and n (if E occurs in all of the n trials); hence, $R(E)$, the relative frequency of E, will always be a number between 0 and 1. The more probable E is, the larger we can expect $R(E)$ to be. Since $R(E)$ always lies between 0 and 1, and $R(E)$ is an approximation to the probability of E, we might also say that the probability of an event is always a number between 0 and 1. In Section 10.2 we shall investigate properties that we would expect probabilities to have, based on certain properties of relative frequencies. We close this section by introducing notation that we shall need in our discussion.

Definition 10.3. *Suppose that A and B are any two events. We shall use $\overline{A}$ to denote the nonoccurrence of A, that is, the event of A not occurring. We shall use (A, B) to denote the event of A and B occurring together, or jointly. The symbol A + B will denote the occurrence of A, or B, or possibly both together; that is, A + B denotes the occurrence of at least one of the events A and B. We use A | B to denote the occurrence of A given that B has already occurred; we can read A | B as "the event A given the event B."*

Example 4. Suppose A is the event of rain, and B is the event of high winds. Then $\overline{A}$ is the nonoccurrence of rain, the event of "no rain" while $\overline{B}$ is the event of "no-high winds." Also, (A, B) is the event of high winds and rain together, while $A + B$ is the event of either rain or high winds (or both rain and high winds). Further, $A \mid B$ is the event of rain when (given that) there are high winds, while $B \mid A$ is the event of high winds when it is raining.

Example 5. Suppose A is the event of passing English, B is the event of taking Russian, and C is the event of being elected to Phi Beta Kappa. Then $(\overline{A}, C)$ is the event of not passing English and being elected to Phi Beta Kappa. And $(B, A + C)$ is the event of both taking Russian and either passing English or being elected to Phi Beta Kappa or passing English and being elected to Phi Beta Kappa. The reader should immediately confirm that $(B, A + C)$ is equivalent to $(B, A) + (B, C)$. We should compare this equivalence with Section 2.6, Exercise 2(d). We can consider two events to be equivalent, or interchangeable, if one event is a necessary and sufficient condition for the other event; that is, if the events always occur and "non-occur" together. Of course, in our discussion, any event may replace any other event to which it is equivalent. For example $(B, A + C)$ can be used in place of $(B, A) + (B, C)$, and conversely.

EXERCISES 10.1

1. Let A be the event of lightning, B be the event of hail, C be the event of high winds, and D be the event of a temperature over 70°. Write out in words each of the following events.

a) $\overline{C}$ b) (A, B) c) $B + D$
d) $A \mid D$ e) $\overline{C} \mid B$ f) $(\overline{A}, \overline{C})$
g) $B \mid (A + C)$ h) $(A, C) + (B, D)$ i) $(\overline{A}, B + \overline{C})$
j) $(A \mid B, C + D)$ k) (A, B, C) l) $A + B + D$
m) $(\overline{A}, B, D)$ n) $(A + B) \mid (C, D)$

2. Verify that the two events in each of the following pairs of events are equivalent; that is, prove that if the first event occurs, then the second also occurs, and that if the second occurs, then the first event also occurs. In each of the following, A, B, and C represent arbitrary events.

a) $(A, B + C)$; $(A, B) + (A, C)$ b) $(A + B, C)$; $(A, C) + (B, C)$

c) A; $\overline{\overline{A}}$ d) $(\overline{A, C})$; $\overline{A} + \overline{C}$

e) $A \mid (B + C)$; $A \mid B + A \mid C$

3. Let A, B, C, and D have the interpretations in Exercise 1. Suppose a survey of 100 summer days is taken and the following table is compiled listing the number of times that the events shown were observed.

Table 10.1

(A, B)	$(\overline{A}, B)$	$(A, \overline{B})$	$(\overline{A}, \overline{B})$
7	2	13	78

a) How many times does A occur during the 100 days? [*Hint: A* occurs if and only if either (A, B) occurs or $(A, \overline{B})$ occurs.]

b) How many times does B occur?

c) How many times does $\overline{B}$ occur?

d) What is the relative frequency of A? of B? of $\overline{B}$?

e) What is the relative frequency of $(A, \overline{B})$?

f) On the basis of Table 10.1, which seems more likely—hail without lightning or lightning without hail?

g) Can any information be obtained about C or D from Table 10.1?

4. Let 1 through 6 each represent the event of rolling that number on one roll of a fair die. Suppose $P(1)$ is the probability of rolling a 1, $P(2)$ is that of rolling a 2, etc. We saw earlier that $P(1) = \frac{1}{6}$. Assuming that the probability of an event is a number between 0 and 1, where an impossible event has probability 0 and a certain event has probability 1, determine each of the following probabilities.

a) $P(5)$ b) $P(5 + \overline{5})$

c) $P(2, 3)$ [*Hint:* Can one roll both a 2 and a 3 on the same roll?]

d) $P(1 + 2)$ [*Hint:* There are 6 possible rolls, each roll equally probable, and a 1 and a 2 are two of these six rolls.]

e) $P(1 + 4 + 5)$ f) $P(\overline{2} + 3)$ g) $P(\overline{2}, \overline{3})$

10.2 THE BASIC RULES OF PROBABILITY

As we saw in Section 10.2, the probability of an event is supposed to be a quantitative measure of the possibility that the event will occur when certain conditions are fulfilled, e.g., a die is rolled or a certain weather pattern exists. If we think of the probability as a measure of the relative frequency with which a certain event will occur, then it is natural to make the probability of an event a number between 0 and 1, the larger the fraction, the greater the likelihood that the event will occur. That is, the greater the probability, the higher the proportion of events to the number of trials. With this view of probability as being in some sense a relative frequency, we state our first rule about probability.

2. The probability of an event E is a number between 0 and 1. The probability of a certain event is 1, while the probability of an impossible event is 0.

Let us suppose that A and B are two possible outcomes of some trial. We shall assume that the trial has been repeated n times and Table 10.2 has been compiled to show the number of occurrences of each of the events under consideration (cf. Section 10.1, Exercise 3). We see that A occurs in conjunction with B a total of p times, A occurs without B a total of q times, etc. Since in any trial, one (and only one) of the events listed in Table 10.2 must occur, we also have $p + q + r + s = n$, the total number of trials.

Table 10.2

(A, B)	$(A, \overline{B})$	$(\overline{A}, B)$	$(\overline{A}, \overline{B})$
p	q	r	s

Since A occurs when and only when either (A, B), or $(A, \overline{B})$ occurs, the total number of times that A occurs is $p + q$; thus $R(A)$, the relative frequency of A, is $(p + q)/n$. Similarly, $R(\overline{A}) = (r + s)/n$. Consequently,

$$R(A) + R(\overline{A}) = (p + q)/n + (r + s)/n$$
$$= (p + q + r + s)/n = n/n = 1.$$

Actually, all that these equations show is that in any trial either A occurs or A does not occur. To repeat the key result of this paragraph:

3. $R(A) + R(\overline{A}) = 1$.

The events $A + B$ and $\big((A, B) + (A, \overline{B}) + (\overline{A}, B)\big)$ are equivalent. For if $A + B$ occurs, then A and B occur together, (A, B), or A occurs without B, $(A, \overline{B})$, or B occurs without A, $(\overline{A}, B)$; similarly, if

$$\big((A, B) + (A, \overline{B}) + (\overline{A}, B)\big)$$

occurs, then either A or B must occur. Table 10.2 shows us then that $A + B$

occurs $p + q + r$ times; hence

$$R(A + B) = (p + q + r)/n.$$

And $R(A)$, as we have seen, is $(p + q)/n$, while $R(B)$, as we may readily compute, is $(p + r)/n$. Adding $R(A)$ to $R(B)$, we obtain

4. $R(A) + R(B) = (p + q)/n + (p + r)/n$
$= (p + q + r)/n + p/n = R(A + B) + p/n.$

We see that p/n is nothing but $R(A, B)$ since (A, B) occurs p times out of n trials; thus (4) becomes

5. $R(A) + R(B) = R(A + B) + R(A, B),$

or

6. $R(A + B) = R(A) + R(B) - R(A, B).$

There are in all $p + q$ trials in which A occurs. Of these $p + q$ trials, B occurs in precisely p (that is, when A and B occur together). Therefore $R(B \mid A)$, the relative frequency of B in those trials in which A occurs, is $p/p + q$. Of course, we can divide both numerator and denominator of this fraction by n without affecting its value, thus obtaining

7. $R(B \mid A) = \dfrac{p/n}{(p + q)/n} .$

But p/n is $R(A, B)$, while $(p + q)/n$ is $R(A)$. Substituting these values in (7), we obtain

8. $R(B \mid A) = \dfrac{R(A, B)}{R(A)}$

or

9. $R(A, B) = R(A)R(B \mid A).$

An entirely similar computation can be used to show

10. $R(A, B) = R(B)R(A \mid B).$

Formulas (3), (6), (9), and (10) form the basis for some of the most important formulas in the theory of probability. Since we expect the relative frequency of an event to approximate more and more closely the true probability of the event as we accumulate trials, we should expect that these formulas also hold for probabilities.

There are certain difficulties in considering a probability as a relative frequency, and there are difficulties in trying to identify any relative frequency with a genuine probability. Most of these difficulties arise from the fact that the improbable can, and sometimes does, happen. Even if the true probability of flipping a "head" is $\frac{1}{2}$, we might flip 1,000,000 consecutive "tails." On the other hand, we might flip a coin 1,000,000 times and get exactly 500,000 heads; then, after the experiment, we may find that the true probability of flipping a head on any toss is but $\frac{1}{10}$. The idea that a relative frequency "approximates" a probability has necessarily been stated informally.

We are in no sense apologizing because problems exist in probability theory, just as we did not apologize about certain fundamental difficulties in set theory. The problems are indeed there, but it is clearly beyond the pretensions of this text to try to state the problems fully or to offer solutions. Nevertheless, what has been said contained a substantial amount of solid mathematics and should have provided the reader with some insight into at least one approach to the basic formulas of probability. We shall assume these formulas as axioms. Specifically, if A and B are events and $P(A)$ represents the probability of A, then:

11. $P(A)$ is a number between 0 and 1, inclusive.

12. $P(A) + P(\overline{A}) = 1$.

13. $P(A + B) = P(A) + P(B) - P(A, B)$.

14. $P(A, B) = P(A)P(B \mid A) = P(B)P(A \mid B)$.

We conclude this section with a number of examples illustrating the use of these formulas.

Example 6. Let 1 through 6 be the events of rolling those particular faces on one roll of a fair die. Then, as we have already seen,

$$P(1) = P(2) = \cdots = P(6) = \tfrac{1}{6};$$

that is, the probability of rolling any one of the faces is $\frac{1}{6}$. Since

$$P(1) = \tfrac{1}{6} \quad \text{and} \quad P(1) + P(\overline{1}) = 1,$$

we have

$$P(\overline{1}) = \tfrac{5}{6};$$

that is, the probability of not rolling a 1 on any particular roll is $\frac{5}{6}$.

Now let us find the probability of rolling either a 3 or a 4, that is, $P(3 + 4)$. By (13) we have

$$P(3 + 4) = P(3) + P(4) - P(3, 4).$$

Now $P(3) = P(4) = \frac{1}{6}$. Since it is quite impossible to roll both a 3 and a 4 on the same roll of one die, $P(3, 4) = 0$; thus

$$P(3 + 4) = \tfrac{1}{6} + \tfrac{1}{6} - 0 = \tfrac{2}{6} = \tfrac{1}{3}.$$

Hence the probability of rolling either a 3 or a 4 on one roll of a fair die is $\tfrac{1}{3}$.

Since simultaneously rolling a 3 and a 4 is impossible, $P(3, 4) = 0$; but by (14),

$$P(3, 4) = P(3)P(4 \mid 3).$$

Also

$$P(3) = \tfrac{1}{6}.$$

Note, however, that $4 \mid 3$ is the event of rolling a 4 given that the roll is a 3. This too is impossible (that is, the roll cannot be a 4 if it is a 3); hence $P(4 \mid 3) = 0$, which is what it must be if equation (14) is to hold in this particular instance.

Example 7. Suppose A and B are events such that $P(A) = \tfrac{3}{4}$ and

$$P(B) = \tfrac{1}{2}.$$

Then we can immediately conclude that $P(A, B)$ is at least $\tfrac{1}{4}$, and hence that it is possible for A and B to occur together (since the nonzero probability indicates that (A, B) is a possible event). We know that

$$P(A + B) = P(A) + P(B) - P(A, B)$$
$$= \tfrac{3}{4} + \tfrac{1}{2} - P(A, B) = \tfrac{5}{4} - P(A, B).$$

Since $P(A, B)$ must be a number less than or equal to 1, $P(A, B)$ must be at least $\tfrac{1}{4}$; otherwise, $\tfrac{5}{4} - P(A, B)$ would be greater than $\tfrac{4}{4} = 1$. We do not have sufficient information to compute precisely what $P(A, B)$ should be. We note, however, that $P(A + B)$ must be at least as large as $P(A)$ since it is at least as likely that $A + B$ will occur as that A will occur. This means that $P(A + B)$ will be greater than or equal to $\tfrac{3}{4}$, which in turn implies that $P(A, B)$ will not exceed $\tfrac{1}{2}$.

Example 8. We shall now derive an expression for $P(A + B + C)$, where A, B, and C are events. The event $A + B + C$ is, of course, the event which occurs if and only if at least one of the events A, B, or C occurs; $A + B + C$ might just as well be written $A + (B + C)$. Then

15. $P(A + B + C) = P(A + (B + C))$
$$= P(A) + P(B + C) - P(A, B + C)$$

by statement (13). The event $(A, B + C)$ is equivalent to, and hence interchangeable with, the event $(A, B) + (A, C)$ (Section 10.1, Exercise 2); hence

$$P(A, B + C) = P((A, B) + (A, C)).$$

By statement (13)

$$P((A, B) + (A, C)) = P(A, B) + P(A, C) - P((A, B), (A, C)).$$

The event $((A, B), (A, C))$ is equivalent to the event (A, B, C); for if $((A, B), (A, C))$ occurs, then A and B occur *and* A and C occur, hence A, B, and C all occur, that is, (A, B, C) occurs. And if (A, B, C) occurs, then (A, B) occurs and (A, C) occurs; hence $((A, B), (A, C))$ occurs. Therefore

$$P((A, B), (A, C)) = P(A, B, C).$$

We thus have

16. $P(A, B + C) = P(A, B) + P(A, C) - P(A, B, C).$

Statement (13) also gives us

17. $P(B + C) = P(B) + P(C) - P(B, C).$

Substituting back into statement (15), we obtain the desired expression for $P(A + B + C)$:

18. $P(A + B + C) = P(A) + P(B) + P(C)$
$$- P(A, B) - P(A, C) - P(B, C) + P(A, B, C).$$

Note that if A, B, and C are events no two of which can occur simultaneously, that is, if (A, B), (A, C), and (B, C)—and thus (A, B, C) as well—are impossible events, then

$$P(A + B + C) = P(A) + P(B) + P(C)$$

since the remaining terms of (18) are then all 0.

EXERCISES 10.2

1. In each of the experiments below, we have probabilities for certain events followed by an event with no assigned probability. Find the least value and the greatest value that the probability of that event can have. If possible, compute its exact probability. (As an example of this type of problem, see Example 7.)

    a) $P(A) = \frac{4}{5}$, $P(B) = \frac{3}{5}$; (A, B)
    b) $P(A, B) = \frac{1}{4}$, $P(A) = \frac{1}{2}$; $B \mid A$
    c) $P(A, B) = \frac{1}{4}$, $P(A) = \frac{1}{2}$; B
    d) $P(A) = 0.4$, $P(B \mid \overline{A}) = 0.3$; $(\overline{A}, B)$
    e) $P(A + B) = 0.5$, $P(A, B) = 0.4$; A

2. Prove that if A, B, and C are events, then

$$P(A, B, C) = P(A)P(B \mid A)P(C \mid A, B).$$

[*Hint:* Write (A, B, C) as $((A, B), C)$ and use (14) to get

$$P((A, B), C) = P(A, B)P(C \mid A, B.)$$

Then apply (14) again to get the desired formula.

3. An event A is said to be *independent* of the event B if $P(A) = P(A \mid B)$. Explain the significance of saying that $P(A) = P(A \mid B)$; in other words, why is it appropriate to say that A and B are independent if this equality is satisfied? What happens to (14) if A and B are events that are independent of each other.

4. Two events A and B are said to be *mutually exclusive* if $P(A, B) = 0$. Explain the significance of $P(A, B) = 0$; that is, tell why the phrase "mutually exclusive" is in fact appropriate. What happens to (13) if A and B are mutually exclusive events? Is it possible for two events to be both mutually exclusive and independent of each other?

5. The probability of rain is 0.1, while the probability of lightning is 0.05. The probability of either rain or lightning (or both) is 0.12.

   a) Find the probability of both rain and lightning.

   b) Find the probability of rain if there is lightning.

   c) Find the probability of lightning if there is rain.

   d) Are lightning and rain mutually exclusive events? independent events?

   e) Find the probability of neither lightning nor rain.

   f) Do we have sufficient facts to compute each of the following probabilities? If so, compute the probabilities; if not, list what added information would enable us to complete the computation.

      i) The probability of rain *or* lightning, but not both

      ii) The probability of lightning without rain

      iii) The probability of either rain or no lightning

10.3 METHODS OF ASSIGNING PROBABILITIES

Statements (11) through (14) give us certain rules for dealing with probabilities once we have them, but the fact is that if we wish to find the probability of an event in the concrete, that is, as a specific number, we must be able to assign numerical probabilities to appropriate events. A man may be highly expert in the theory of yachting, but unless he has a ship, he will never sail. Likewise, if we are to make our probability theory serve us in the concrete, we must have tools for assigning probabilities to events.

We have already touched on one method of assigning probabilities, namely, the relative-frequency approach. The relative-frequency approach is the experimental approach to probability; we accumulate data through actual observation or experimentation, then assign probabilities in accordance with the observed data. For example, if the Bureau of the Census finds that 18,000,000 out of 200,000,000 Americans have red hair, we can assign the probability $18/200 = 0.09$ to the event that a randomly chosen American has red hair.

The Classical Approach. Another approach to assigning probabilities is the so-called *classical approach.* We used this approach at the very beginning of this chapter to determine that the probability of flipping heads with a fair coin is $\frac{1}{2}$ and that of rolling a 1 with a fair die is $\frac{1}{6}$. More generally, let us suppose this situation: there are precisely n possible outcomes of a given trial with the added conditions that (1) no two of these events can occur together and (2) there is no reason to suspect that any one of the events is more probable than any other event. For the sake of simplicity, we shall denote these n events by 1 through n. Since the given trial must have one of the events as its outcome, the chance that at least one of the events will occur is certainty; that is,

19. $P(1 + 2 + \cdots + n) = 1.$

Next we use the Principle of Finite Induction to prove the following proposition.

Proposition 1. *If* $1, 2, \ldots, n - 1, n$ *are* n *events such that no two can occur together, then*

$$P(1 + 2 + \cdots + (n - 1) + n) = P(1) + P(2) + \cdots + P(\mathrm{n} - 1) + P(n);$$

that is, the probability that at least one (and hence exactly one since no two can occur together) of these n *events will occur is equal to the sum of the probabilities of the individual events.*

Proof. When $n = 2$, we know from statement (13) that

$$P(1 + 2) = P(1) + P(2) - P(1, 2).$$

Since $(1, 2)$ is impossible, $P(1, 2) = 0$, hence $P(1 + 2) = P(1) + P(2)$. Thus Proposition 1 is true when $n = 2$. We now show that whenever the proposition is true for n, it is also true for $n + 1$. Assume that Proposition 1 holds for any set of n events such that no two of the events can occur together, and let $1, 2, \ldots, n, n + 1$ be a set of $n + 1$ events which also has the property that no two of these events can occur together. Then

$$P(1 + 2 + \cdots + n + (n + 1)) = P((1 + 2 + \cdots + n) + (n + 1)),$$

which in turn, by statement (13) is equal to

$$P(1 + 2 + \cdots + n) + P(n + 1) - P((1 + 2 + \cdots + n), n + 1).$$

Now $1, 2, \ldots, n$ are n events, no two of which can occur together; hence by the induction assumption

$$P(1 + 2 + \cdots + n) = P(1) + P(2) + \cdots + P(n).$$

Also $\big((1 + 2 + \cdots + n), n + 1\big)$ is the event of $n + 1$ occurring in conjunction with at least one of the events $1, 2, \ldots, n$, which is impossible; thus

$$P\big((1 + 2 + \cdots + n), n + 1\big) = 0.$$

We therefore have

$$P\big(1 + 2 + \cdots + n + (n + 1)\big) \\ = P(1) + P(2) + \cdots + P(n) + P(n + 1),$$

which is Proposition 1 for $n + 1$. By the Principle of Finite Induction, then, Proposition 1 is in fact true for any positive integer n greater than or equal to 2. For $n = 1$, Proposition 1 reduces to the trivial case $P(1) = P(1)$.

Applying Proposition 1 to statement (19), we obtain

20. $P(1) + P(2) + \cdots + P(n) = 1.$

By assumption, $P(1) = P(2) = \cdots = P(n)$; thus we have

21. $nP(1) = 1 \qquad$ or $\qquad P(1) = 1/n = P(2) = \cdots = P(n).$

To summarize, if some trial has n possible outcomes, each outcome as probable as any other and no two outcomes able to occur simultaneously, then a probability of $1/n$ should be assigned to each of the n outcomes.

Example 9. A roll of a fair die can have six possible outcomes $1, 2, \ldots, 6$. No two of these outcomes can occur simultaneously, hence a probability of $\frac{1}{6}$ should be assigned to each.

Example 10. Suppose 5 people, A, B, C, D, and E, are in a room. One person gets up and leaves. What is the probability that the person is A? Since we are given no reason to suspect that one person's leaving is any more probable than another person's leaving, and since if only 1 person leaves then it is impossible to have 2 people gone simultaneously, each person should be assigned probability $\frac{1}{5}$; hence $P(A)$, the probability that it was A who left, is $\frac{1}{5}$.

Example 11. Suppose we wish to find the probability of getting an odd number on rolling a fair die. In order to obtain an odd number, the event

$1 + 3 + 5$ must occur. Thus, Proposition 1 tells us that

$$P(\text{odd number}) = P(1 + 3 + 5) = P(1) + P(3) + P(5),$$

and Example 9 showed us that

$$P(1) + P(3) + P(5) = \tfrac{1}{6} + \tfrac{1}{6} + \tfrac{1}{6} = \tfrac{1}{2}.$$

Example 12. Suppose M is the number of ways in which 5 different cards can be drawn from a standard deck of 52 cards; that is, M is the number of 5-card poker hands. Since no one hand is more probable that any other hand, the probability to be assigned to each individual hand is $1/M$.

We might ask now what is the probability of getting a hand which contains only hearts. Let us suppose that there are K hands which contain all hearts. Then to each hand containing all hearts, we can assign a distinct number between 1 and K, and we shall identify the hand with that number. Thus, in order to get a hand with all hearts, it is necessary and sufficient that the event $1 + 2 + \cdots + K$ occur. Since no two distinct hands can occur simultaneously, we can apply Proposition 1 to obtain

$$P(1 + 2 + \cdots + K) = P(1) + P(2) + \cdots + P(K).$$

But $P(1) = P(2) = \cdots = P(K) = 1/M$ since the probability of any particular hand is $1/M$. Thus the probability we are looking for is

$$P(1 + 2 + \cdots + K) = K(1/M) = K/M.$$

We should keep in mind that even if some particular trial can have n outcomes, and even if it is also true that no two of these outcomes can occur together, it is still not necessarily true that each of the events has the same probability. This point is illustrated in the following example.

Example 13. Suppose we roll two fair dice. Then we may obtain a number between 2 (if 1 appears on both dice) and 12 (if 6 appears on both dice). It is, of course, impossible to roll two distinct totals simultaneously, thus the events $2, 3, 4, \ldots, 11, 12$ are mutually exclusive in pairs. Nevertheless, these events do not have the same probabilities. For example, there is only one way to roll a 2, but there are six distinct ways to roll a 7 (namely, a 1 on the first die and a 6 on the second, a 6 on the first and a 1 on the second, a 2 on the first and a 5 on the second, etc.); hence we have good reason for suspecting that a 7 is six times as probable as a 2.

The question now arises: How can we find an appropriate set of events which are equiprobable and which give the possible outcomes of a fair roll of two fair dice? Any particular roll of the two dice can be expressed as an ordered pair (n, m), where n tells us what happened on the first die, and m tells us what happened on the second die; thus $(1, 6)$ would represent a 1 on the first die and a 6 on the second die. The numbers n and m can take on values between 1 and 6, inclusive. That is, any particular roll of the two dice can be represented by a unique element of the Cartesian product

$$\{1, 2, \ldots, 6\} \times \{1, 2, \ldots, 6\};$$

moreover, each element of this Cartesian product represents a unique roll of the two dice. There are 36 elements of

$$\{1, 2, \ldots, 6\} \times \{1, 2, \ldots, 6\},$$

hence there are precisely 36 possible rolls with the two dice. Of these 36 possible rolls, no one roll is any more probable than any other roll. Therefore to each one of these 36 rolls, we assign the probability $\frac{1}{36}$. We can now compute the probability of rolling a 7 as follows: In order to have a 7 it is necessary and sufficient that at least one (and hence exactly one) of the events $(3, 4)$, $(4, 3)$, $(2, 5)$, $(5, 2)$, $(1, 6)$, $(6, 1)$ occur. Keep in mind that each event is determined by its order as well as its makeup; thus $(3, 4)$ does not indicate rolling a 3 and 4 with both dice simultaneously, but rather rolling a 3 on the first die and a 4 on the second die. Then

$$P(7) = P\big((3, 4) + (4, 3) + (2, 5) + (5, 2) + (1, 6) + (6, 1)\big)$$
$$= P(3, 4) + P(4, 3) + \cdots + P(6, 1) = 6(\tfrac{1}{36}) = \tfrac{1}{6}.$$

Note that we could also compute $P(3, 4)$ using (14). For

$$P(3, 4) = P(3 \text{ on the first die})P(4 \text{ on the second die} \mid 3 \text{ on the first die}).$$

We know that $P(3 \text{ on the first die}) = \frac{1}{6}$, and since what we roll on the first die in no way determines what we roll on the second die, we also have $P(4 \text{ on the second die} \mid 3 \text{ on the first die}) = P(4 \text{ on the second die}) = \frac{1}{6}$. Thus $P(3, 4) = \frac{1}{6} \cdot \frac{1}{6} = \frac{1}{36}$.

If it is known that a certain trial can have exactly n possible outcomes which are mutually exclusive in pairs (that is, no two can occur together) and which are all equally likely, then we can assign probability $1/n$ to each of these events. But, of course, it is not always easy to express the possible outcomes of a trial in this form (the form we used for the roll of two dice in Example 13); and even if we can visualize a means to so express the possible events, we are often left with the problem of finding their exact number. For example, in Example 12 we computed the probability as K/M for a poker hand containing only hearts. There still remains the not insignificant problem of evaluating K and M. In point of fact, the computation of these numbers is not really a problem in probability, but an exercise in bookkeeping. For the sake of completeness and to enable the reader to obtain purely numerical solutions for a number of probability problems, we shall present several methods of computation in Section 10.4.

EXERCISES 10.3

1. Find the probability of rolling each of the following on a roll of two fair dice:
    a) 2   b) 4   c) 6
    d) 9   e) 11.

2. In each of the following, a certain probability is asked for. First find a set of equiprobable, pairwise mutually exclusive events which adequately represent the possible outcomes of the appropriate trial. Denote the number of these events by n if you cannot compute their number. Next find the desired probability in terms of n.

    Illustration: Six people, A, B, C, D, E, and F, are in a room. Two leave; what is the probability that the two are A and B? Let n be the number of combinations in which two people can leave the room. The probability that the two people are A and B is $1/n$ since A and B leaving together is just one of the n equiprobable and pairwise mutually exclusive ways in which two people can leave.

    Also compute the desired probability, if you can, by any suitable means. In the illustration, we can let (A, B) denote the event of A leaving *first* and B leaving second, while (B, A) will denote B leaving first with A leaving second. Then A and B will both leave if and only if the event $(A, B) + (B, A)$ occurs. Now
    $$P((A, B) + (B, A)) = P(A, B) + P(B, A).$$
    We find that
    $$P(A, B) = P(A \text{ leaving first})P(B \text{ leaving} \mid A \text{ has left first}) = \tfrac{1}{5} \cdot \tfrac{1}{4}$$
    (cf. Example 10). Now we see that
    $$P(B \text{ leaving} \mid A \text{ has left}) = \tfrac{1}{4}$$
    since after A has left there are but 4 people left. Similarly,
    $$P(B, A) = \tfrac{1}{5} \cdot \tfrac{1}{4};$$
    hence the desired probability is
    $$2 \cdot \tfrac{1}{20} = \tfrac{1}{10}.$$

    There does not necessarily have to be only one correct method of solution for any given problem.

a) Slips numbered from 1 through 6 are in an urn. Someone reaches in and draws two of the slips. What is the probability that they are the slips numbered 3 and 4?

b) There are five balls in an urn, two red and three blue. Someone reaches into the urn and pulls out two balls. What is the probability that both are red?

c) A man flips a fair coin seven consecutive times. What is the probability of getting heads on all seven flips?

d) Ten people, five women and five men, are in a room. Four people leave the room. What is the probability that all four are women?

e) What probability should be assigned to one particular bridge hand (that is, to one particular deal of 13 cards from a standard deck of 52 cards). What is the probability of getting a bridge hand containing 12 spades and 1 card of another suit?

f) Three fair dice are rolled together. What is the probability that their faces will total 4?

g) Ten people each roll ten fair dice. What is the probability that at least nine of the people (i.e., nine or ten of them) get a 1?

h) There are four balls in an urn, two red and two blue. Someone picks one of the balls, records its color, and replaces it in the urn. This process is repeated four times. What is the probability that a red ball is drawn twice and a blue ball twice?

3. With a certain die the probability of rolling a 3 is twice that of getting any other face. What is the probability of rolling a 3 with this die? Suppose this die is paired with a fair die and the two are rolled together. What is the probability of getting a 7 with the pair?

4. A certain coin has only $\frac{1}{3}$ as much chance of coming up heads as tails. What is the probability of flipping heads with the coin? What is the probability of getting tails on two consecutive flips?

10.4 SAMPLING

Suppose five people, A, B, C, D, and E, are candidates for two positions. How many ways can the positions be filled? As it stands now, we cannot answer this question because we do not have sufficient information to completely determine the answer. First, we have not been told whether or not the same person might fill both positions. Second, the positions might be such that the selection of A to fill position 1 and B to fill position 2 would be essentially the same as having B fill position 1 and A fill position 2 (for example, if we were selecting cochairmen of a committee); on the other hand, A in position 1 and B in position 2 might be a completely different arrangement from one in which B holds position 1 and A holds position 2 (for example, if we were choosing the chairman and the secretary of a committee).

Consider another situation. Suppose there are five distinct balls, A, B, C, D, and E, in an urn, and we are to make one selection from the urn three times. In how many ways can this be done? Once again the problem is underdetermined. First, we do not know whether the order in which we make the selections is to be taken into account. That is, we are not told whether drawing A first, B second, and C third is to be considered the same as drawing C first, A second, and B third; we get the same balls, but in a completely different order. Also, we are not told whether or not we replace the ball after each draw. If we replace the ball, then we could draw A three times, which would, of course, be impossible if we did not return each draw to the urn before drawing again.

Let us suppose that in general we wish to select m objects from a set of n objects* and we want to know in how many ways the selection can be made. We have to answer the questions:

1) Is the order in which the objects are selected to be taken into account?
2) Do we replace an object after we have selected it?

We now consider some of the various cases that can arise.

CASE 1. We are to select m objects from n objects. Each object will be set aside after it is selected, that is, a selected object will not be returned to the pool of objects. Order will be taken into account so that two possible selections that consist of the same m objects are still considered different so long as the order of the selection is different.

We can represent any particular selection of m objects from the n objects (of course, m can be at most as large as n) as an ordered m-tuple $(a_1, a_2, \ldots, a_m)$, where a_1 is the object selected first, a_2 the object selected second, etc. Since there is no replacement of objects, each of the "coordinates" in the m-tuple must be different. Also any ordered m-tuple formed of m different objects from the set of n objects represents one specific way to select m objects without replacement and with order taken into account. Therefore, if we can determine the number of such ordered m-tuples, we will also know the number of ways the m objects can be selected in accordance with the given conditions.

Starting from scratch, we would have n ways of filling the first coordinate of such an m-tuple, one for each of the n objects. We would then have only $n - 1$ ways of filling the second coordinate, one for each of the remaining $n - 1$ objects after the first object was removed. Using an argument almost identical to that of Proposition 1, Chapter 7, we see that there are then $n(n - 1)$ ways of filling the first two coordinates of the m-tuple. Since now only $n - 2$ objects remain, there are $n - 2$ ways

---

* This general process of selection is known as *sampling*.

of filling the third coordinate, or $n(n - 1)(n - 2)$ ways of filling the first three coordinates. Continuing in like fashion, we proceed until we see that there are exactly $n - m + 1$ ways of filling the last coordinate, or $n(n - 1)(n - 2) \cdots (n - m + 2)(n - m + 1)$ ways of filling all the coordinates; that is, the total number of ordered m-tuples that can be formed from the n objects without repetition in any coordinate is

22. $_nP_m = n(n - 1)(n - 2) \cdots (n - m + 1).$

A particular choice of m objects from a total of n objects, taking order into account and without replacement, is called a *permutation of n objects taken m at a time*. The symbol $_nP_m$ represents the number of permutations of n objects taken m at a time.

Example 14. Suppose 5 people, A, B, C, D, and E, are in a room. We might want to know in how many ways all these people can file out of the room one at a time. This would be the number of permutations of 5 things taken 5 at a time, that is, $_5P_5 = 5 \cdot 4 \cdot 3 \cdot 2 \cdot 1 = 120$.

If we want to know in how many ways 2 people could leave the room, taking the order in which they left into account, we would have

$$_5P_2 = 5 \cdot 4 = 20.$$

Applying this finding to probability theory, we can assign a probability of $\frac{1}{20}$ to any particular manner in which 2 people can leave the room and taking the order of their exits into account. We can also assign a probability of $\frac{1}{120}$ to any particular manner in which the 5 people can file out of the room.

CASE 2. We are to select m objects from n objects. Once selected, an object remains out of the n-collection, but order of selection will not be taken into account; that is, we are now interested only in the objects selected and not in the order in which selection occurs.

Let $a_1, \ldots, a_m$ be a possible selection of m objects. The subscripts here merely distinguish the objects, they do not indicate the order in which the objects were picked. We first ask how many permutations of n objects taken m at a time correspond to this one possible selection of m objects; that is, how many permutations select $a_1, \ldots, a_m$. Since each of the permutations will select exactly the same objects, the only way in which these permutations will differ from one another is in the order in which they select the objects; there will be one permutation for each of the ways in which we can order the m objects. But the number of ways in which m objects can be ordered is the number of permutations of m

things taken m at a time (that is, the number of ways in which we can select the m objects one at a time, taking the order of selection into account). Since

$$_mP_m = m(m-1)(m-2) \cdots 2 \cdot 1,$$

this is the number of permutations of n things taken m at a time which select the same m objects.

Suppose now that we look at all $_nP_m$ permutations of n things taken m at a time. Given two permutations p and q, we shall say that p is *equivalent to* q if p and q select the same m objects. Then "is equivalent to" defines an equivalence relation on the set of permutations. As we have seen above, each equivalence class of permutations contains $_mP_m$ elements. Moreover, the number of equivalence classes is exactly the number of ways of selecting m objects from n objects without taking order into account (and without replacement), and this number is $_nP_m/_mP_m$. (This argument is quite similar to the following: Suppose 15 people are split into 5 committees such that everyone is on one and only one committee and each committee has the same number of members. Then each committee will have $\frac{15}{5} = 3$ members.)

The number of ways of selecting m objects from n objects without replacement or regard for order is called the number of *combinations of n things taken m at a time* and will be denoted by $C(n, m)$;* thus we have shown that

$$C(n, m) = \frac{_nP_m}{_mP_m} = \frac{n(n-1) \cdots (n-m+1)}{m(m-1) \cdots 2 \cdot 1}.$$

Example 15. Suppose we have a group of 20 people. In how many ways can a committee of 5 be selected from this group assuming that the order of selection is unimportant? The answer is

$$C(20, 5) = \frac{_{20}P_5}{_5P_5} = \frac{(20 \cdot 19 \cdot 18 \cdot 17 \cdot 16)}{(5 \cdot 4 \cdot 3 \cdot 2 \cdot 1)} = 15{,}504.$$

We might also ask: How many committees of 5 are there which include a particular person, say person A, as a member? If one particular person is assumed to be a member of the committee, then that leaves 4 places to be filled from 19 remaining people. Thus the number of committees with A as a member is

$$C(19, 4) = \frac{19 \cdot 18 \cdot 17 \cdot 16}{4 \cdot 3 \cdot 2 \cdot 1} = 3876.$$

----

* The notation $\binom{n}{m}$ is also widely used.

If some particular committee of 5 is selected from the 20 people, the probability that the committee will have A as a member is $3876/15{,}504 = \frac{1}{4}$. For any individual committee has exactly probability $1/15{,}504$ of being selected and there are 3876 of the possible committees with A as a member.

CASE 3. We are to select m times from n objects. An object is to be replaced after it is selected and the order of selection is to be taken into account.

Here we have almost the same conditions as those in Case 1 except that now the objects are replaced. We can again represent any particular selection by an ordered m-tuple $(a_1, \ldots, a_m)$, but since there is replacement, there can be repetition in the coordinates. The total number of ordered m-tuples that can be formed from n objects, allowing for repetition in the coordinates, is

$$n \cdot n \cdots (m \text{ times}) \cdot n, = n^m,$$

that is, the number of elements in the Cartesian product of $\{1, 2, \ldots, n\}$ with itself m times.

Example 16. Suppose an urn contains 3 red balls and 4 blue balls. A ball is drawn from the urn, the color is recorded, and the ball is replaced. The procedure is repeated 8 times. How many possible ways are there for the string of 8 draws to come out, provided that the order of selection is taken into account? On each draw we have only *two* possibilities, red and blue, even though there are 7 balls. Thus we are asking how many ways are there to select 8 times from 2 objects (note that m can be greater than n here since there is replacement) with replacement and taking order into account. This result is $2^8 = 256$. Note that the probability of any one such selection is not necessarily $\frac{1}{256}$ since there is good reason to suppose that the selections are not equally likely; namely, there are more blue balls than red balls, and hence a string containing all blue balls, for example, would be more probable than a string of all red balls.

We shall not consider a case that calls for replacement but is not concerned with order. We close this section by giving several examples of the application to probability theory of the computational devices developed thus far.

Example 17. Six slips in an urn are numbered 1 through 6. Someone reaches into the urn and draws out three of the slips. What is the probability that the slips drawn are 4, 5, and 6? From the wording of the problem it is reasonable to assume that our sampling is one without replacement and without regard to order (cf. Case 2). The total number of ways in which three slips can be drawn from the urn under the given conditions is $C(6, 3)$. Since no one way of drawing three slips is intrinsically more probable than any other way, and since we cannot make simultaneously two distinct draws of three slips, the probability to be assigned to any one draw of three slips is $1/C(6, 3)$. Since

$$C(6, 3) = \frac{6 \cdot 5 \cdot 4}{3 \cdot 2 \cdot 1} = 20,$$

this probability is $\frac{1}{20}$. Since 4, 5, 6 is precisely one way of selecting three slips, the probability of 4, 5, 6 is $\frac{1}{20}$.

Example 18. A fair coin is flipped 7 times. What is the probability of obtaining exactly 4 heads in the 7 flips? This situation corresponds to Case 3. The total number of ways in which the 7 flips can occur (taking order into account) is 2^7. Since heads and tails are equally likely, each of the 2^7 ways is equally likely; hence the probability of any one particular string of heads and tails is $\frac{1}{2}^7 = \frac{1}{128}$. Now we ask how many of these 2^7 strings give us exactly 4 heads. Thinking of each (ordered) string of 7 flips as an ordered 7-tuple, if a string is to contain exactly 4 heads, it must have heads in exactly 4 coordinates. The problem thus reduces to the question: In how many ways can we choose 4 coordinates from the 7 coordinates without replacement (since each coordinate is considered as an independent entity) and without regard to order (since each of the coordinates is to be filled by precisely the same things, heads)? The answer to this latter question is $C(7, 4)$. The probability then of obtaining precisely 4 heads in the 7 flips is $C(7, 4)(\frac{1}{128}) = \frac{35}{128}$.

We shall consider this type of situation again in a more general context in Section 10.5.

Example 19. Five people, A, B, C, D, and E, are in a room. They file out one by one. What is the probability that A leaves first, and E leaves last? There are, as we have already seen, 120 ways for the five people to leave the room. We must now answer the question: In how many of these ways is A first and E last? Let us consider the various ways in which the five people can leave the room as ordered 5-tuples; those 5-tuples having A in the first coordinate and E in the last coordinate are precisely those which correspond to the event of A leaving first and E last. The middle three coordinates of such 5-tuples must be filled by B, C, and D. The number of ways in which B, C, and D can fill the middle coordinates is $_3P_3$; thus there are $_3P_3 = 6$ ways for the five people to leave the room with A first and E last. The probability then that the five people will file out with A first and E last is $6(\frac{1}{120}) = \frac{1}{20}$.

EXERCISES 10.4

1. Evaluate numerically each of the following:
   a) $_4P_3$ b) $C(4, 3)$
   c) $C(10, 8)$ d) $C(156, 1)$
   e) $_9P_8$ f) $_7P_0$

2. If n is a positive integer, n *factorial* (written $n!$) is defined as the product of all positive integers less than or equal to n; that is,

$$n! = n(n - 1)(n - 2) \cdots 2 \cdot 1.$$

   For example,

$$4! = 4 \cdot 3 \cdot 2 \cdot 1.$$

   For the sake of convenience $0!$ is defined as 1. The factorial notation is often used to simplify the formulas for $_nP_m$ and $C(n, m)$. Prove

   a) $_nP_m = \dfrac{n!}{(n - m)!}$

   b) $C(n, m) = \dfrac{n!}{(n - m)!m!}$.

3. Compute the number of ways in which
   a) 4 objects can be selected from a total of 8 objects without replacement and without regard to order of selection;
   b) 4 objects can be selected from a total of 8 objects with replacement and with regard to the order of selection;
   c) 4 objects can be selected from a total of 8 objects without replacement and with regard to the order of selection;
   d) a fair die can be rolled 6 consecutive times;
   e) 6 fair dice can be rolled all at once;
   f) a committee of 5 people might be formed from a group of 10 people;
   g) 5 distinct positions can be filled from a group of 10 people, assuming no person can hold more than 1 position;
   h) 5 distinct positions can be filled from a group of 10 people, assuming any person can hold more than 1 position.

4. The following should utilize the information obtained in Exercise 3.
   a) What is the probability of rolling 6 consecutive 1's with a fair die?
   b) If a committee of 5 is to be formed from 10 people, what is the probability that the committee will contain 3 particular people, say A, B, and C?
   c) What is the probability of getting exactly two 3's in six rolls of a fair die?
   d) Eight slips numbered 1 through 8 are in an urn. A selection of four slips is made from the urn without replacement. What is the probability that the four slips drawn are 6, 7, 2, and 3?
   e) Eight slips are in an urn as in (d). Four are drawn out one at a time. What is the probability that slip 1 will be drawn first and slip 4 will be drawn fourth?

10.5 SOME SPECIAL PROBABILITIES

In this section we consider probabilities in certain specialized, though not uncommon, situations.

SITUATION 1. There are n objects of two types: p of Type 1, and $n - p$ of Type 2. We are to select m objects from the n objects without replacement or regard to order of selection. What is the probability of getting exactly q objects of Type 1? Before proceeding to solve this problem, we present several illustrations of the situation.

Example 20. There are 12 red balls and 15 blue balls in an urn. We are to select 10 balls from the urn without replacement and without regard to order. What is the probability of getting exactly 4 red balls? Note that in this case, the total number of objects $n = 12 + 15 = 27$. There are 12 Type 1 objects (red balls) and 15 Type 2 objects (blue, or nonred, balls). In this case, $m = 10$ and $q = 4$. It should be apparent that in order for a problem of this type to be sensible, m can be at most as large as n, p can be at most as large as n, and q can be at most as large as p.

Example 21. A paper bag contains eight \$1 bills, seven \$5 bills, and six \$10 bills. A man reaches into the bag and draws out four of the bills (without replacement or regard to order). What is the probability that he gets exactly two of the \$10 bills? Here $n = 8 + 7 + 6 = 21$, $p = 6$ (since a Type 1 object is a \$10 bill), $m = 4$, and $q = 2$.

Within the terms of Situation 1, there are $C(n, m)$ ways of selecting the m objects, hence the probability of any particular selection (considering all n objects to be distinct entities) is $1/C(n, m)$. We now ask how many of these $C(n, m)$ ways of selecting will give us exactly q objects of Type 1. We consider a selection of m objects to be a set of m objects, hence we know that the selection will have precisely q objects of Type 1 (and consequently $m - q$ objects of Type 2) if precisely q of the m objects in the set are objects of Type 1. Since there are exactly p objects of Type 1 to choose from, there are $C(p, q)$ ways to select the Type 1 objects. Similarly, there are $C(n - p, m - q)$ ways in which the $m - q$ objects of Type 2 in the set can be chosen. Consequently, there is a total of

$$C(p, q)C(n - p, m - q)$$

ways in which the selection of m objects can contain precisely q objects of Type 1. We therefore conclude that the probability of getting q objects of Type 1, that is, the probability originally requested, is

23. $\dfrac{C(p, q)C(n - p, m - q)}{C(n, m)}$.

Example 22. In Example 20, (23) would be

$$\frac{C(12, 4)C(15, 6)}{C(27, 10)}.$$

In Example 21, (23) would be

$$\frac{C(6, 2)C(15, 2)}{C(21, 4)}.$$

The numerical evaluation of these quantities is left to the determined reader.

SITUATION 2. Any particular trial of a certain experiment must have one and only one of two outcomes A and B. We assume $P(A) = p$; then

$$P(B) = 1 - P(A) = 1 - p.$$

The experiment is performed n times. What is the probability that A will occur in exactly m of the n trials? Again we can illustrate the situation.

Example 23. A fair die is rolled 30 times. What is the probability that 15 of those rolls will be either 4 or 5? We denote event A by $4 + 5$ and event B by $\overline{4 + 5}$. On any roll, either A or B must occur, but both cannot occur together. Thus $P(A) = \frac{1}{6} + \frac{1}{6} = \frac{1}{3}$, while $P(B)1 - \frac{1}{3} = \frac{2}{3}$. Of course $n = 30$ and $m = 15$.

Example 24. The probability of a forest fire in the United States on any given day is 0.08. What is the probability of having a forest fire on exactly 3 days in a month of 30 days? Let A be the event of a forest fire on a given day and B be $\overline{A}$; then $P(A) = 0.08$ and $P(B) = 0.92$. Here $n = 30$ and $m = 3$.

The results of an ordered string of n trials can be represented by an ordered n-tuple in which the result of the first trial is recorded in the first coordinate, the result of the second trial in the second coordinate, etc. An important thing to notice is that the trials are independent of one another; that is, the results of one trial do not influence the results of any other trial. This is implicit in the assumption that on any given trial

$$P(A) = p \quad \text{and} \quad P(B) = 1 - p;$$

that is, these probabilities are unaffected by any previous results. We can therefore conclude that any string which contains m events A is just as likely as any other string which also contains m events A; that is, the order in which the m events A occur is irrelevant to the probability of the occurrence of the string since no trial is affected by what happens before or after it. The probability of a string containing m events A depends then only on the fact that it contains m events A. We now compute the probability of such a string. In particular, we compute the probability of the string in which A occurs in the first m trials and B occurs thereafter. What we wish to find is

$$P(A_1, A_2, \ldots, A_m, B_{m+1}, \ldots, B_n),$$

where A_i denotes A occurring on the ith trial (or A being in the ith coordinate of the ordered n-tuple), and B_j indicates that B occurs on the jth trial. Using a repeated application of (14) together with the fact that the trials are independent, we obtain

24. $P(A_1, \ldots, A_m, B_{m+1}, \ldots, B_n)$
$= P(A_1)P(A_2) \ldots P(A_m)P(B_{m+1}) \ldots P(B_n) = p^m(1 - p)^{n-m}.$

In sum, the probability of the occurrence of some one particular sequence of n trials in which A occurs m times is $p^m(1 - p)^{n-m}$. The question now is: How many such sequences are there?

In order to have such a sequence, we must fill m coordinates of an ordered m-tuple with A's. Thus the question reduces to: In how many ways can we select m coordinates from n coordinates (without replacement and without regard to order since all the m coordinates will be filled in exactly the same way)? This, of course, is $C(n, m)$.

Since we have $C(n, m)$ events—the $C(n, m)$ strings each containing m A's—which are mutually exclusive in pairs and each of which has a probability of $p^m(1 - p)^{n-m}$, we can apply Proposition 1 to obtain the probability of getting exactly m events A in n trials:

25. $C(n, m)p^m(1 - p)^{n-m}.$

Example 25. In Example 23, (25) becomes

$$C(30, 15)(\tfrac{1}{3})^{15}(\tfrac{2}{3})^{15},$$

while in Example 24, we have

$$C(30, 3)(0.08)^3(0.92)^{27}.$$

The computation of these quantities is left to the determined reader.

SITUATION 3. Suppose that some event A can occur only in conjunction with one of several events $B_1, B_2, \ldots, B_n$, and that the $B_1, \ldots, B_n$ are mutually exclusive in pairs. Suppose we are also given the information that A has occurred. Then since A occurs only when one of the $B_1, \ldots, B_n$

occurs, one of the $B_1, \ldots, B_n$ must have occurred. What is the probability that it was B_1 that occurred with A; that is, what is $P(B_1 \mid A)$?

From (14) we have

$$P(A, B_1) = P(A)P(B_1 \mid A) = P(B_1)P(A \mid B_1).$$

From this we obtain

26. $P(B_1 \mid A) = \dfrac{P(B_1)P(A \mid B_1)}{P(A)}.$

Since A must occur in conjunction with one of $B_1, \ldots, B_n$, the event A is equivalent to the event

$$(A, B_1) + (A, B_2) + \cdots + (A, B_n).$$

Since the events $B_1, \ldots, B_n$ are mutually exclusive in pairs, the events $(A, B_1), \ldots, (A, B_n)$ are also mutually exclusive in pairs (for if, for example, (A, B_1) and (A, B_2) occurred together, then B_1 and B_2 would also occur together, which, by hypothesis, is impossible). Then

$$P(A) = P((A, B_1) + \cdots + (A, B_n))$$
$$= P(A, B_1) + P(A, B_2) + \cdots + P(A, B_n).$$

Applying (14), we obtain

27. $P(A) = P(B_1)P(A \mid B_1) + P(B_2)P(A \mid B_2) + \cdots + P(B_n)P(A \mid B_n).$

Hence we have

28. $P(B_1 \mid A)$
$$= \frac{P(B_1)P(A \mid B_1)}{P(B_1)P(A \mid B_1) + P(B_2)P(A \mid B_2) + \cdots + P(B_n)P(A \mid B_n)}.$$

Statement (28) is known as *Bayes' Theorem.*

Example 26. Urn X contains 4 red and 3 blue balls, while urn Y contains 5 red and 6 blue balls, and urn Z contains 4 red and 4 blue balls. A man reaches into one of the urns and pulls out 3 balls, all of which are found to be blue. What is the probability that the man drew from urn X? Here A should be interpreted as drawing 3 blue balls; B_1, B_2, and B_3 represent choosing urns X, Y, or Z, respectively, from which to draw. Since we have no reason to feel that the man making the selection will be attracted to one urn more than another, we may assume that

$$P(B_1) = P(B_2) = P(B_3) = \tfrac{1}{3}.$$

We see that $P(A \mid B_1)$ is the probability of drawing 3 blues from urn X

if 3 balls are selected from the urn. This type of probability was handled in Situation 1 and can be found to be

$$\frac{C(3, 3)C(4, 0)}{C(7, 3)} = \frac{1}{C(7, 3)}.$$

Similarly, $P(A \mid B_2)$ and $P(A \mid B_3)$ are readily computable as

$$\frac{C(6, 3)}{C(11, 3)} \quad \text{and} \quad \frac{C(4, 3)}{C(8, 3)},$$

respectively. In this case, (28) becomes

$$P(B_1 \mid A) = \frac{\frac{1}{3}(1/C(7, 3))}{\frac{1}{3}(1/C(7, 3)) + \frac{1}{3}(C(6, 3)/C(11, 3)) + \frac{1}{3}(C(4, 3)/C(8, 3))}.$$

Example 27. A man has heard that one of three slot machines is defective and pays the jackpot twice as often as it should, but he is not sure which of the three machines it is. He plays one of the machines and hits the jackpot. What is the probability he has picked the defective machine? Let A be hitting the jackpot, B_1 be picking the defective machine, and B_2 be $\overline{B}_1$. Let p be the probability of hitting a jackpot on a nondefective machine. Then $P(A \mid B_1) = 2p$, while $P(A \mid B_2) = p$. Since all the machines are equally likely prospects, but there are two nondefective machines and only one defective machine, we have $P(B_1) = \frac{1}{3}$ and $P(B_2) = \frac{2}{3}$. Thus we have

$$P(B_1 \mid A) \text{ (the desired probability)} = \frac{\frac{1}{3}(2p)}{\frac{1}{3}(2p) + \frac{2}{3}(p)} = \frac{1}{2}.$$

Before applying any of the formulas derived for any of the three situations treated in this section, the reader must be certain that he is indeed dealing with one of these situations. The following example illustrates a case which looks as though it could be handled by Bayes' Theorem, but which actually does not fulfill the necessary conditions.

Example 28. In a certain part of the world rain always occurs together with hail, lightning, or high winds. Letting A be rain, B_1, B_2, and B_3 be hail, lightning, and high winds, respectively, we are given that

$$P(B_1) = 0.09, \quad P(B_2) = 0.67, \quad P(B_3) = 0.23,$$
$$P(A \mid B_1) = 0.96, \quad P(A \mid B_2) = 0.78, \quad P(A \mid B_3) = 0.5.$$

Even though all the terms necessary for completing (28) are given, (28) cannot be used for finding $P(B_1 \mid A)$, since the events B_1, B_2, and B_3 are not mutually exclusive in pairs (that is, it is quite possible to have hail and high winds together).

EXERCISES 10.5

1. An urn contains 8 blue balls and 10 green balls. Someone reaches into the urn and draws out 7 balls. What is the probability that exactly 5 of these balls will be green?

2. An urn contains 8 blue balls and 10 green balls. Someone reaches into the urn, draws out 1 ball, records the color of the ball, and then replaces it in the urn. The process is repeated 7 times. What is the probability that a green ball is drawn on exactly 5 of those draws?

3. Twelve people are in a room: 7 men and 5 women. Three people leave the room. What is the probability that 2 who left are women and 1 is a man?

4. What is the probability of getting a bridge hand which contains exactly 6 spades?

5. A card is drawn from a standard deck of 52 cards, the suit is recorded, and the card is returned to the deck. The process is repeated 12 times. What is the probability of drawing a spade exactly 5 of those times?

6. Urn A contains 4 red balls and 7 blue balls, while urn B contains 6 red balls and 14 blue balls. A selection of 2 balls is made from one of the urns and it is found that 1 red ball and 1 blue ball have been drawn. What is the probability that the selection was made from urn B?

7. Brand X controls 30% of the market; Brand Y, 60%; and Brand Z, 10%. Three housewives compare notes and find that they all use the same brand. What is the probability that all three use Brand X?

8. What is the probability of getting a bridge hand containing 5 spades, 4 clubs, 2 hearts, and 2 diamonds? [*Hint:* Generalize (23) slightly.]

9. The probability of a fire in a private home on any given day is 0.0001. A certain city contains 8000 private homes. What is the probability that there will be 3 or fewer fires (that is, 3, 2, 1, or 0) on any given day?

10. A shipment of 100 widgets contains 3 defective ones. A sample of 4 widgets is taken from the shipment and tested. What is the probability that *exactly* 1 of the defective widgets will be found? that *at least* 1 of the defective widgets will be found?

11. One of 4 pennies is known to always fall showing "heads" when flipped, while the other 3 pennies are undoctored. One of the pennies is selected and flipped 5 times and each time, "heads" occurs. What is the probability that the weighted penny has been chosen?

12. Urn A contains 3 red and 4 blue balls, urn B contains 4 red and 5 blue balls, and urn C contains 5 red and 6 blue balls. One ball is selected from each of the three urns. What is the probability that 3 blue balls are drawn? that *exactly* 1 blue ball is drawn? that *at least* 1 blue ball is drawn?

chapter eleven

An Elementary Geometry

11.1 THE PROJECTIVE PLANE

Thus far, although we have discussed extensively the axiomatic nature of mathematics, we have not developed any mathematical system from a purely axiomatic point of view. We know that the systems we have studied, for example, logic and set theory, can be approached axiomatically, but we ourselves did not really use such an approach. When studying probability, we did arrive at certain statements which we accepted as being axioms; yet even then, the study of probability was not purely axiomatic in the sense that everything we learned about probability rested ultimately on those axioms.

The axiomatic techniques necessary to do justice to many of the topics of this text are simply beyond the scope of the book. Moreover, the approaches we *did* use, though not strictly axiomatic, probably furnished better insight into the mathematical procedure than a rigorous axiomatic development could have provided. The picture of mathematics we have seen so far is fairly accurate, despite its shortcomings. But we should attempt to understand at least one genuine axiomatic system before concluding this study.

One of the simplest and yet richest of axiomatic systems is that of the projective plane. The fact that some mathematical system is based on undefined concepts and unproved statements does not mean that the mathematical system is not motivated or inspired by the "real world." Indeed, projective geometry has its roots in the very manner in which we see. Before we consider the axioms for a projective plane, we shall first motivate those axioms so that we shall find it natural to study a projective plane.

It is well known that when we look down straight railroad tracks, the parallel rails seem to meet in the distance. This phenomenon of seeing things in perspective, was not incorporated into Western art until the

Renaissance. Before that time the artists' depictions were "flat" and strictly two-dimensional. They lacked the depth or third dimension that we perceive in nature. With the incorporation of perspective into art came an interest in studying the principles involved in the phenomenon of perspective. For example, consider the lampshade of Fig. 11.1. The bottom edge of the lampshade may be a perfect circle, but how it looks to us will depend on the angle from which we view it. If we view it from the side, the bottom edge may appear to be a straight line or an ellipse. In fact, it will appear in its true circular shape only if we look at it from directly above or directly below. The point is that the appearance of a geometrical object is determined by our viewpoint. The appearance of an object at some given angle is a question of great importance to the artist, and hence, it is not surprising that extensive investigation on this matter was conducted even while modern mathematics was still in its infancy and that much of this investigation was carried out by artists.

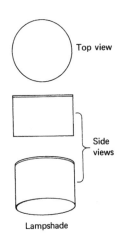

Top view

Side
views

Lampshade

We may try to express axiomatically at least some of the properties of our perspective vision. We restrict ourselves for simplicity to two dimensions, that is, to how we view things on a flat surface.

Axiom 1.   Any two points lie on precisely one line.

Axiom 2.   Any two lines lie on precisely one point.

Fig. 11.1

Axiom 1 embodies the idea that two points determine a line. Axiom 2 corresponds to that property of our vision which makes any two lines seem to intersect. The relationship "lie on" is used in both axioms and, like the terms "line" and "point," is left undefined. Note that we do not have to know what these terms mean in order to be able to draw conclusions about them in the abstract.

In order to ensure that we have something nontrivial to consider, and in particular, to be sure that we have more than one line, we add

Axiom 3.   There are at least three points which do not all lie on the same line

Axioms 1, 2, and 3 may appear at first glance to be so elementary that it may seem they cannot take us very far. But the fact is that a considerable amount of mathematics rests essentially on these three axioms. Even during the course of this short chapter we shall encounter problem based on these axioms which remain unsolved to this day.

When the axioms that determine the mathematical system are formulated, we must throw away our preconceived opinions about what must be true in the system that the axioms define. We must essentially rely on the axioms to prove what we feel to be true. We may indeed be led to some particular theorem by observing the example which inspired the axioms, but we must prove that theorem either from the axioms alone or from other statements derived ultimately from the axioms. We might remark that if we cannot prove a certain statement but still feel that the statement must be true in order for the mathematical system to be what we want it to be, then we can add the statement as another axiom.*

Now let us consider some examples of specific objects which can be interpreted as satisfying Axioms 1, 2, and 3, but which have nothing to do with our mode of vision, the original inspiration for the three axioms.

Example 1. The objects under consideration will be the three vertices of a triangle. If we call each vertex a point and each pair of vertices a line, and interpret "lie on" as meaning "are common to" or "have in common," then the three vertices satisfy Axioms 1, 2, and 3. In this case the axioms read:

Axiom 1. Any two vertices are common to a unique pair of vertices.

Axiom 2. Any two pairs of vertices have a unique vertex in common.

Axiom 3. The three vertices are not all common to a single pair of vertices.

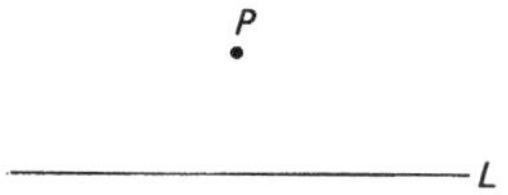

Fig. 11.2

Example 2. Let L be any line in the Euclidean plane and let P be any point of the plane which does not lie on L (see Fig. 11.2). We shall interpret any point of the plane which lies on L as being a point and P will also be a point. Suppose P_1 and P_2 are distinct points. We interpret "lie on" in Axioms 1, 2, and 3 as meaning "determines," or "is common to." We define the unique line on which P_1 and P_2 lie as follows.

If P_1 or P_2 is P, then the line on which P_1 and P_2 lie will just be $\{P_1, P_2\}$. If both P_1 and P_2 are on L, then this line will be L.

It is left as an exercise to verify that Axioms 1, 2, and 3 are satisfied by this example.

---

* Provided, of course, that the statement does not contradict, that is, is not logically inconsistent with, the axioms we have already assumed.

Example 3. Consider the integers 1, 2, 3, 4, 5, 6, and 7. We shall call each integer a point and our lines will be the following collections of three integers each: {1, 2, 3}, {3, 4, 5}, {1, 5, 6}, {1, 4, 7}, {3, 6, 7}, {2, 5, 7}, and {2, 4, 6}. "Lie on" again will mean "determine." A graphic representation of the lines and points and their relationships is given in Fig. 11.3. The reader should verify that this system satisfies Axioms 1, 2, and 3.

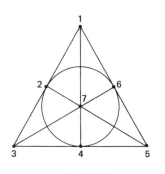

Fig. 11.3

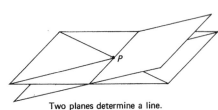

Two planes determine a line.
Two lines determine a plane.

Fig. 11.4

Example 4. Let R^3 be Euclidean 3-space, that is, the space of Euclidean solid geometry, and let P be a point of R^3. Any two planes through P intersect in a straight line (Fig. 11.4). We shall call each *line* through P a point and each *plane* through P a line. "Lie on" will be interpreted by "determine." Once again Axioms 1, 2, and 3 (page 220) are satisfied since any two ordinary lines through P (*points* in our new terminology) determine (lie on) a unique plane (line), while any two planes determine a unique line (that is, two lines lie on one point). Axiom 3 is trivially satisfied.

In Examples 1 and 2 we have lines that contain only two points. Since two-point lines are not particularly interesting, and since the possibility of two-point lines could complicate our discussion by forcing us to consider "special cases" within a given proposition, we shall for convenience add the following assumption as an axiom:

Axiom 4. Each line lies on at least three points.

Our discussion will also be simplified by making the following assumption:

Axiom 5. A point P lies on a line L if and only if L lies on P.

Definition 11.1. Any system of lines and points which satisfies Axioms 1 2, 3, 4, and 5 is said to be a projective plane.

We shall investigate some of the fundamental properties of projective planes in Section 11.2.

EXERCISES 11.1

1. Verify that Examples 2 and 3 do in fact satisfy Axioms 1, 2, and 3. Note that Example 3 gives a projective plane while Example 2 does not.

2. Prove that Axioms 1 through 4 are independent; that is, show that no one of the axioms can be derived as a logical consequence of the others. This can be shown by finding examples of systems which satisfy all the axioms except one; that one axiom will then be known to be independent of the others. For if it were not independent (if the one axiom could be proved from the others) then any system that satisfied the other axioms would also have to satisfy that one. For example, in Example 1 we have a system that satisfies Axioms 1, 2, and 3, but not Axiom 4. We thus conclude that Axiom 4 cannot be derived from Axioms 1, 2, and 3.

3. Consider $\{1, 2, 3, 4\}$. Suppose the elements of this set are to be interpreted as points and certain subsets of this set are to be interpreted as lines with "lie on" meaning "determine." Prove that no such system formed from $\{1, 2, 3, 4\}$ can be a projective plane. Try to determine whether such a system can satisfy Axioms 1, 2, and 3.

4. Try to formulate a set of axioms analogous to Axioms 1 through 5 for *projective 3-space*. These axioms should embody the properties of our 3-dimensional visual perception. One axiom, for example, will be: Any two planes lie on precisely one line.

11.2 BASIC PROPERTIES OF THE PROJECTIVE PLANE

Throughout this section we shall assume that we are dealing with a projective plane. If A and B are points, then A and B lie on a unique line in accordance with Axiom 1. We shall denote this line by AB.

Proposition 1. *If A, B, and C are distinct points which lie on a line L, then $L = AB = BC = AC$.*

Proof. By Axiom 1 there is one and only one line AB such that A and B lie on AB. Since A and B both lie on L, it must be that $AB = L$. A similar argument shows that $BC = L$ and $AC = L$. Therefore

$$AB = BC = AC = L.$$

Proposition 2. *There are at least three lines all of which do not lie on the same point.*

Proof. By Axiom 3 we have three points A, B, and C which do not all lie on the same line. We now prove that the lines AB, BC, and AC do not all lie on the same point. We begin by showing that all these lines are actually distinct from one another. If $AB = BC$, then C must lie on AB, which would imply that A, B, and C all lie on the same line; but such an arrangement contradicts the assumption that A, B, and C do not all lie on the same line. Therefore $AB \neq BC$. Similar reasoning shows that $AC \neq BC$ and $AB \neq AC$. Suppose AB, BC, and AC all lie on the same point D. Since AB and BC both lie on D and AB and BC also both lie on B (by Axiom 5), it follows that $D = B$ since, by Axiom 2, AB and BC can lie on only one point. But then AC also lies on B, hence AC lies on A, B, and C. But then A, B, and C all lie on AC, contradicting the assumption that A, B, and C do not all lie on the same line. Therefore AB, BC, and CD do not all lie on the same point.

The following proposition shows that Axiom 3 and Proposition 2 are equivalent if all axioms except (3) are assumed. That is, if we make Proposition 2 into an axiom and assume Axioms 1, 2, 4, and 5 as well, then Axiom 3 can be proved as a proposition.* Of course, we have already shown that Proposition 2 can be proved from Axioms 1 through 5. This proposition illustrates the point that a mathematical system may have several different (although equivalent) axiomatic foundations.

Proposition 3. *If we assume Axioms 1, 2, and 5, then Proposition 2 is true if and only if Axiom 3 is true.*

Proof. We have shown that if Axiom 3 is true, then Proposition 2 is true. Suppose now that Proposition 2 is true; then there are lines L_1, L_2, and L_3 which do not all lie on the same point. By Axiom 2, L_1 and L_2 both lie on some point A, L_2 and L_3 both lie on some point B, and L_1 and L_3 both lie on some point C. Suppose $A = B$. Then $C \neq A$, or L_1, L_2, and L_3 would all lie on the same point. But since $A = B$, it follows that L_1 and L_3 both lie on $A(=B)$ and C. Therefore A and C both lie on L_1 and L_3; hence $L_1 = AC = L_3$, contradicting the fact that L_1 and L_3 are distinct lines. Therefore we have shown that $A \neq B$. Similar reasoning shows that $B \neq C$ and $A \neq C$. Now if A, B, and C all lie on the same line L_4, then, since $L_1 = AC$ and $L_3 = BC$, we have $L_4 = AC = BC$ (Proposition 1) $= L_1 = L_3$, another contradiction of the fact that L_1 and L_3 are distinct lines. Therefore A, B, and C cannot lie on the same line. We have produced three points which do not all lie on the same line; hence Axiom 3 is satisfied and the proof of Proposition 3 is complete.

Proposition 4. *Each point lies on at least three lines.*

Proof. Let P be any point (Fig. 11.5). By Axiom 3 we know that there is at least one other point P'. Again using Axiom 3 we can choose some point

---

* Actually, Axiom 4 was not used in the proof of Proposition 2 and is not required to prove Proposition 3.

Q which does not lie on PP'. Point P cannot lie on $P'Q$, for if it did, then PP' would equal $P'Q$, which would imply that Q lies on PP', thus contradicting the choice of Q. But by Axiom 4, $P'Q$ lies on at least three points, hence there is some point Q' distinct from both P' and Q which lies on $P'Q$. It is easy to see then that P lies on each of the lines $P'P$, PQ, and PQ' and that no two of these lines are the same.

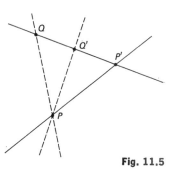

Fig. 11.5

Proposition 5. *If Axioms 1, 2, 3, and 5, are assumed, then Proposition 4 is satisfied if and only if Axiom 4 is satisfied.*

Proof. We have already proved that Proposition 4 holds if Axiom 4 holds (along with the other axioms, of course). Assume now that Proposition 4 is true and let L be any line. There is at least one point P which does not lie on L. By assumption there are at least three distinct lines L_1, L_2, and L_3 which lie on P. By Axiom 2, L and L_1 lie on a unique point P_1, L and L_2 lie on a unique point P_2, and L and L_3 lie on a unique point P_3. Since L_1, L_2, and L_3 are distinct lines, P_1, P_2, and P_3 are distinct points. But all of these points lie on L, hence L lies on at least three points.

We know from Propositions 3 and 5 that Axioms 3 and 4 are interchangeable with the statements of Propositions 2 and 4, respectively; that is, we might have used Proposition 2 (or Proposition 4) as an axiom and proved Axiom 3 (or Axiom 4) instead of having done things the way we did them. Actually, Propositions 3 and 5 have far deeper and more significant implications for they enable us to say that the terms "line" and "point" are essentially interchangeable and that whatever statement is true about a projective plane remains true when "line" and "point" are substituted for each other. We now list Axioms 1 through 5 with the terms *"line"* and *"point"* interchanged.

Axiom 1'. *Any two lines lie on precisely one point.*

Axiom 2'. *Any two points lie on precisely one line.*

Axiom 3'. *There are at least three lines which do not all lie on the same point.*

Axiom 4'. *Each point lies on at least three lines.*

Axiom 5'. *A line L lies on a point P if and only if P lies on L.*

This set of axioms is equivalent to the original set of axioms; that is, we can prove Axioms 1 through 5 by means of Axioms 1' through 5', and vice versa. Observe that Axiom 1′ is the same as Axiom 2, while Axiom 2′ is the same as Axiom 1. Axiom 5′ is essentially the same as Axiom 5 even though *line* and *point* have been interchanged. Axiom 3′ is simply Proposition 2, and Axiom 4′ is Proposition 4. From Proposition 5 we have that Axioms 1 through 5 are equivalent to Axioms 1, 2, 3, and 5, together with Proposition 4. From Proposition 3 we have that Axioms 1, 2, 3, and 5, together with Proposition 4, are equivalent to Axioms 1, 2, and 5, together with Propositions 2 and 4. Therefore Axioms 1 through 5 are equivalent to Axioms 1, 2, and 5, together with Propositions 2 and 4.

Definition 11.2. *If S is any statement about a projective plane, then the statement formed from S by interchanging the terms line and point in S is called the dual statement of S.*

We thus see that Axiom 1′ and Axiom 2 are dual statements, as are Axioms 2′ and 1. Axiom 5 is its own dual, that is, Axiom 5 remains essentially unchanged after "line" and "point" are interchanged; such a statement is said to be *self-dual*. The importance of the equivalence of Axioms 1 through 5 and their dual statements Axioms 1′ through 5′ rests in the fact that if any statement is true about a projective plane, then its dual statement is also true. Thus, whenever we prove any statement about a projective plane, we generally obtain another statement free of charge, the dual of the statement we prove. (We say "generally" not because the dual of a true statement about the projective plane is ever false, but because sometimes the dual turns out to be the same statement, as, for example, in the case of Axioms 5 and 5′; hence the dual does not really give us any new information.) This interchangeability of *"line"* and *"point"* is known as the *Principle of Duality*. Using the Principle of Duality alone we can make the following statements:

Proposition 6.

a) *If there are n points in a projective plane, then there are n lines. (That is, if the statement "There are n points" is true, then its dual "There are n lines" is also true.)*
b) *If there are n lines, then there are n points.*
c) *If each point lies on m lines, then each line lies on m points.*
d) *If each line lies on m points, then each point lies on m lines.*

Note that (a) and (b) are dual statements, as are (c) and (d).

Proposition 7. *Each line lies on the same number of points as any other line.*

Proof. Let L_1 and L_2 be any two lines. Then L_1 and L_2 lie on the same point P (see Fig. 11.6). By Proposition 4 there is at least one other line L_3 that lies on P. By Axiom 4 there are at least three points that lie on

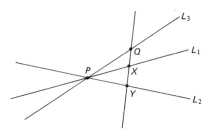

Fig. 11.6

L_3, hence we can choose Q, a point lying on L_3, but different from P. It is clear that Q cannot lie on either L_1 or L_2. Let X be any point that lies on L_1. Then QX and L_2 lie on the same point Y. We pair the point X of L_1 with the point Y of L_2. In this manner, each point X that lies on L_1 is paired with exactly one point Y that lies on L_2; moreover, each point that lies on L_2 is paired with exactly one point that lies on L_1 since if Y is any point lying on L_2, then some point X lies on both L_1 and QY, hence Y is paired with X (and X alone). We have therefore found a one-one and onto function from the set of points lying on L_1 onto the set of points lying on L_2. Lines L_1 and L_2 therefore lie on the same number of points. (For more about the idea of two sets having the same number of elements, the reader is referred to Chapter 6 and Proposition 7, Chapter 7. Certain details of the proof of Proposition 7 are asked for in Exercise Set 11.2.)

As a corollary of Proposition 7, we have the following dual statement:

Proposition 8. *Any point lies on exactly as many lines as any other point.*

If each line lies on an infinite number of points, then obviously the number of points in the projective plane is infinite. We may ask, however, how many points are there if each line lies on precisely n points.

Proposition 9. *If each line lies on n points, then there are exactly $n^2 - n + 1$ points.*

Proof. Let L be any line and P be any point that does not lie on L. There are n distinct lines of the type PX, where X is some point that lies on L, that is, there is one line for each point of L. Each of these lines lies on n points and P is the only point which any two of these lines have in common. Any point lies on at least one of these lines since if Y is any point, then PY and L lie on some point X; hence Y lies on PX. If all lines of the form PX had no points in common, we would have precisely n^2 points. But all these lines lie on P; hence to avoid counting P n times, we must take $n - 1$ (one for each of the n lines except one) from n^2. This procedure leaves us with a total of $n^2 - (n - 1) = n^2 - n + 1$ points.

Corollary (*The dual of Proposition 9*). *If each point lies on* n *lines, then there are exactly* $n^2 - n + 1$ *lines.*

From Proposition 6 we conclude:

Proposition 10. *If each line lies on* n *points, then there are* $n^2 - n + 1$ *points and* $n^2 - n + 1$ *lines.*

We thus conclude that there can be no projective plane with 5 or 6 points since neither 5 nor 6 is of the form $n^2 - n + 1$ for any positive integer n. If $n = 3$, then $n^2 - n + 1 = 7$, hence there might be a projective plane that has 7 points and 7 lines with 3 points on each line. Such a plane was given in Example 3. We might ask whether, given a positive integer of the form $n^2 - n + 1$, we can discover a projective plane with that many points. The answer to this question is still one of the great unsolved problems of mathematics. Another unsolved problem is the following: If there is at least one projective plane containing $n^2 - n + 1$ points, then how many "distinct" planes are there? (We shall not elaborate on what we mean by "distinct.") Only partial answers to both these questions have been found, and complete answers appear to be a long way off.

EXERCISES 11.2

1. Prove that the pairing described in the proof of Proposition 7 pairs each point of L_1 with precisely one point of L_2; that is, prove that $Y = Y'$ if Y and Y' are points of L_1 which are paired in the manner described in the proof of Proposition 7 with some point X of L_2. Prove also that each point of L_2 is paired with one and only one point of L_1.

2. It is possible to have a projective plane of 13 points and 13 lines since $13 = 4^2 - 4 + 1$. Construct such a projective plane using as points the integers from 1 through 13. The lines of this plane should be taken as sets of four points each. Be sure that any two lines intersect in one and only one point and that any two points determine exactly one line.

3. Is the Principle of Duality true for ordinary Euclidean plane geometry? Specifically, can "line" and "point" always be interchanged in propositions that are true in plane geometry so as always to give a true statement? *Consider:* Given a point P which does not lie on a line L, there is precisely one line which lies on P and does not lie on any points common to L.

4. In a projective plane we have no notion of length or line segments (unless we add more axioms); nevertheless, we can talk about certain geometric configurations. A *triangle* is defined as a system of 3 points which do not all lie on the same line, together with 3 lines, each line lying on 2 points and each point lying on 2 lines (Fig. 11.7). What is the configuration which is dual to a triangle; that is, interchange *line* and *point* in the definition of a triangle and what kind of configuration do you get?

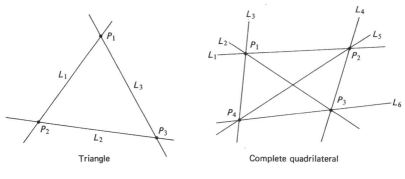

Triangle   Complete quadrilateral

Fig. 11.7   **Fig. 11.8**

We define a *complete quadrilateral* as a system of 4 points and 6 lines such that each point lies on precisely 3 of the lines and each line lies on precisely 2 of the points (Fig. 11.8). What is the configuration dual to a complete quadrilateral? The dual configuration is called *a complete quadrangle*. Draw a diagram to illustrate such a quadrangle.

5. Can there be a projective plane in which each line lies on an infinite number of points (see Example 4)? Can there be a projective plane with 9 points? If each line in a projective plane contains $n + 1$ points, then the projective plane is said to be of *order* n. Prove that any projective plane of order n contains $n^2 + n + 1$ points.

11.3 THE AFFINE PLANE

In a projective plane any two lines lie on some one point, that is, informally, any two lines intersect. In Euclidean plane geometry, it is not true that any two lines intersect since it is quite possible for two lines to be parallel. In a projective plane we have no such thing as parallel lines. If we wished to set up an axiom system for a geometry in which there was a parallel postulate such as the one in Euclidean geometry, then we could use the following axioms.

Axiom A.   *Any two points lie on precisely one line.*

Axiom B.   *Two lines lie on at most one point.*

Axiom C.   *If a point P does not lie on a line L, then there is precisely one line L' which lies on P such that L and L' do not lie on any common point.*

Axiom D.   *There are at least three points which do not all lie on the same line.*

Axiom E.   *A point P lies on a line L if and only if L lies on P.*

Definition 11.3. *A system satisfying Axioms A through E is said to be an affine plane. If π′ is an affine plane and L and L′ are lines in π′ such that L and L′ do not lie on a common point, or such that L = L′, then L is said to be parallel to L′. (Note that we have defined any line to be parallel to itself.)*

The following proposition shows that there is an affine plane associated with every projective plane.

Proposition 11. *Let π be a projective plane and L be any line in π. We form a new structure π′ from π as follows: The points of π′ will be all of the points of π except for those which lie on L. The lines of π′ will be the lines of π except for L. A line L′ of π′ will lie on any point of π′ that it originally lay on in π, and any point of π′ will lie on any line of π′ that it originally lay on in π. We shall use the same notation to represent a line or point that is in π′ as we used when the line or point was in π. Then π′ is an affine plane.*

Proof. We must verify Axioms A through E for π′.

Axiom A: If P and P' are any two points of π′, then P and P' are also points of π. Therefore P and P' lie on a unique line PP' of π; moreover, since P and P' are in π′, it is impossible that $PP' = L$ (the line that was removed along with all points which lie on it). Therefore PP' is a line in π′ and is the only line on which P and P' both lie.

Axiom B: If H and H' are any two lines of π′, then H and H' are two lines of π. Since H and H' can lie on at most one point in π, H and H' can lie on at most one point in π′.

Axiom C: Suppose H is any line in π′ and P is any point of π′ which does not lie on H. Then H and P are also in π. Now H and L both lie on some point Q. Then Q lies on both H and PQ. Both H and PQ are lines in π′, and P lies on PQ. But there is no point of π′ which lies on both H and PQ. For if there were, this point would have to be Q and Q is not a point of π′. Therefore PQ and H are parallel. It is left as an exercise to prove that PQ is the only line in π′ which lies on P and is parallel to H.

The verification of Axioms D and E are also left as exercises.

Corollary. *If the projective plane π has $n^2 - n + 1$ points, then the affine plane π′ derived from π has $(n - 1)^2$ points and $n(n - 1)$ lines.*

Proof. Since we are removing from π all n points that lie on one line, we are left with

$$(n^2 - n + 1) - n = n^2 - 2n + 1 = (n - 1)^2$$

points. Since π also has $n^2 - n + 1$ lines and we are removing one of them,

we are left with
$$n^2 - n = n(n - 1)$$
lines.

The fact that an affine plane can contain a number of points different from its number of lines tells us at once that we do not have a Principle of Duality for affine planes.

We shall soon show that not only is there an affine plane associated with each projective plane, but there is a projective plane associated with every affine plane. Before proving this important result, however, we must prove certain preliminary results.

Proposition 12. *Any two lines in an affine plane lie on the same number of points.*

Proof. Let L and L' be any two lines of an affine plane π'. Let P be any point that lies on L but not on L', and let P' be any point that lies on L' but not on L. We shall pair P with P'. Suppose X is any point lying on L. Then there is a unique line H_X that lies on X and is parallel to PP' (Fig. 11.9). Now H_X must lie on some point Y of L'. For if H_X were parallel to L', then PP' and L' would be two distinct lines lying on P' and parallel to H_X, a contradiction of Axiom C. We shall pair X with Y. It is left as an exercise to prove that this pairing establishes that L and L' lie on the same number of points.

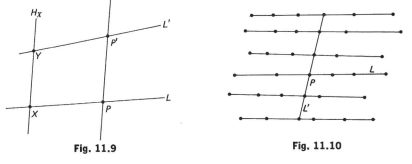

Fig. 11.9   **Fig. 11.10**

Proposition 13. *Let π' be any affine plane in which each line lies on n points. Then there are precisely n^2 points, $n^2 + n$ lines, and given any line L, n lines of π' which are parallel to L. (See Fig. 11.10.)*

Proof. Suppose L is any line of π'. Let L' be any line that lies on some point P of L, but $L \neq L'$. Line L' also lies on n points. Lying on each of these points is a unique line parallel to L and all these lines are distinct. Moreover, if H is any line parallel to L, then H will lie on some point of L'; if it did not, then L and L' would both be lines lying on P and parallel to H, contradicting Axiom C. There are therefore n lines parallel to L, one for each of the n points of L'. Now every point of π' lies on one (and only one) of these lines parallel to L. For if X is any point of π', there is a line parallel to L which lies on X. Each point of π' lies on one of these n lines; each of these lines lie on n points, and no two of these lines lie on the same point. We therefore conclude that there is a total of n^2 points in π'.

We compute the number of lines in π' as follows: Let L be any line and L' be any line parallel to L, but $L \neq L'$. For each point X of L and each point Y of L', we have a distinct line XY (Fig. 11.11). There are n^2 such lines. The only lines left uncounted after all lines of the type XY have been formed are those parallel to L (Exercise 8). There are, however, n such lines. Thus there are in all $n^2 + n$ lines.

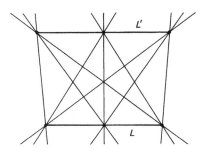

Fig. 11.11

Definition 11.4. *The relationship "is parallel to" defines an equivalence relation on the set of lines of any affine plane (Exercise 3). If L is a line of an affine plane π', then the equivalence class of L is said to be the parallel class of L.*

Corollary. *If π' is an affine plane such that each line of π' lies on n points, then each parallel class contains n lines and there are $n + 1$ parallel classes.*

Proof. That there are n lines in each parallel class follows at once from Definition 11.4 and Proposition 13. Since there are $n^2 + n$ lines and these are partitioned into parallel classes of n members each, there are

$$(n^2 + n)/n = n + 1$$

parallel classes.

Proposition 14. *Let π' be any affine plane. Let S be any set having the same number of elements as π' has parallel classes, and such that no element of S is a line or point of π'. (Thus there will be $n + 1$ elements of S if each line of π' lies on n points.) We select one line from each parallel class of π' and to each line we "add" a point of S in the following sense: If L is one of the selected lines, we choose a unique point Q of S and define Q to lie on L and L*

to lie on Q; each point that originally lay on L continues to lie on L. Every element of S is to be added to one and only one of the selected lines; this is possible because S has the same number of elements as the set of selected lines. We now add the element Q of S which was added to L to each member of the parallel class of L in the same fashion. (Note that now any two members of the parallel class of L both lie on Q.) We shall also let S be a line and say that each element of S lies on S and S lies on each of its elements. The augmented lines of π′ together with S form the lines of the new structure π; the points of π are the points of π′ together with the elements of S. Then π is a projective plane. The verification that Axioms 1 through 5 are satisfied by π is straightforward though somewhat tedious. The details are left to the reader.

Propositions 11 and 14 essentially state that if we start with a projective plane, then we can form an affine plane by omitting one line; and if we start with an affine plane, we can form a projective plane by adding a line. We can therefore draw conclusions about affine planes by studying projective planes since any affine plane can be considered as part of a projective plane. In truth, it is often much easier to study projective planes than affine planes because in projective planes we have the Principle of Duality, plus the simplifying condition that any two lines intersect. In an affine plane two lines might be either parallel or intersecting. We shall not develop this point, or the geometry of projective planes, any further in this text; however, for a further discussion of some of the topics presented in this chapter, the interested reader might see a charming and not difficult book by Harold Dorwart, *The Geometry of Incidence.*[*]

EXERCISES 11.3

1. Verify Axioms D and E in Proposition 11.

2. Prove that the pairing described in Proposition 12 suffices to show that L and L' both lie on the same number of points. In particular, show that each point of L is paired with one and only one point of L' and each point of L' is paired with one and only one point of L.

3. Prove that "is parallel to" defines an equivalence relation on the set of lines in any affine plane. For the notion of an equivalence relation see Section 8.2.

4. Prove Proposition 14.

5. Suppose that each line in an affine plane lies on n points. How many lines lie on any given point? Do this problem in two ways. First, compute this number directly using what we have proved about affine planes. Second, compute this number using what we know about projective planes. That is,

---

[*] Harold Dorwart, *The Geometry of Incidence*. Englewood Cliffs, N.J.: Prentice Hall, 1966.

there is associated with any affine plane π' a projective plane π which "contains" π'. In the associated projective plane, how many lines lie on any given point (cf. Proposition 6)? Why can we then say that the answer is the same for π'?

6. Use the information gathered in Exercise 5 to recompute the number of points in an affine plane in which each line lies on n points.

7. If n is an integer greater than or equal to 2, can we be sure that there is always an affine plane in which each line lies on n points? [*Hint:* If we could be sure, what could we say about projective planes? Can we say it?]

8. In the part of the proof of Proposition 13 that shows that π' has $n^2 + n$ lines, prove that if L'' is any line not parallel to L, then L'' is not parallel to L'. That is, $L'' = XY$ for some point X lying on L and some point Y lying on L'.

Conclusion

12.1 WHERE SHALL WE GO FROM HERE?

This book has presented many of the fundamental concepts of modern mathematics. Now we might ask: Where shall we go from here? We could of course develop further the mathematical systems that we only began to study in this text. There is much more to set theory and logic than what we have seen here. For example, we might try to "coordinatize" projective planes, that is, assign coordinates to the points of a projective plane much as we give coordinates to the points of the usual coordinate plane of analytic geometry. In working on such a development, we would find that the axioms for a projective plane are rich in algebraic implications.

We might also develop material in this text by "specializing" it. For example, we could try to add axioms to the axioms for an affine plane until we had an axiom system for Euclidean geometry. Once we had defined the integers, we could proceed to define the rational numbers, then the real numbers. Once we obtained the real numbers we could develop the calculus or pursue other significant investigations.

We might also "generalize" some of the concepts we have studied. For example, we have found that addition and multiplication of integers have certain properties. We can abstract these properties and use them in various combinations, possibly adding other properties which seem appropriate as well, to obtain a rich variety of algebraic structures.

We might also change certain of the axioms defining some system and investigate what effect the change produces. For example, we might change Axiom C in the definition of an affine plane so that more than one line through a point outside a given line L can be parallel to L. Such a change would give the beginnings of a truly non-Euclidean geometry.

Although we have tried to motivate by intuitive arguments much of what we have studied in this text, we might throw our intuition away and try to devise systems which, despite their running counter to our conditioned ideas of what should be true, could be a significant contribution to knowledge. For example, our development of logic has been based on the premise that there are essentially two truth values. However, logics have been developed which have three or more truth values. These multi-valued logics have proved useful from a philosophical and a practical viewpoint.

In sum then, there are many ways to develop the ideas which this book has introduced all too briefly. There is one other path to mathematical knowledge, in fact to knowledge of any kind—creating one's own ideas and concepts. Most of what the reader finds in this book was undreamt of not very long ago. We cannot know what forms mathematics will assume in the future but we can predict that the mathematics of tomorrow will be shaped by those with the ability and the desire to think creatively.

12.2 MATHEMATICS AS A SCIENCE

Almost anything that people agree to call a science is more than a mere collection of data. Simple random facts pertaining to some general area such as physics, chemistry, or economics do not constitute a science. A science is something organized, systematized, axiomatized, if you prefer.

Consider physics for example. Everyone before Newton observed that if someone dropped an object, that object fell. The astronomer Tycho Brahe had collected vast data on the motion of the planets but all Brahe's data did not constitute a science of planetary motion. It was Johannes Kepler who actually came up with a set of planetary-motion laws which not only explained Brahe's data but also enabled astronomers to predict what other planets would do. Kepler's laws transformed Brahe's data into a body of scientific knowledge. Then Newton developed his calculus and laws of motion which enabled him to derive Kepler's laws of planetary motion as only one of many important consequences.

Let us consider the sequence of events. First, a general phenomenon was observed, the action of gravity on objects. Then many data were compiled on the motion of certain planets. Next, a set of mathematical equations was developed which related the large quantity of previously unrelated data. Then a powerful tool was developed, the calculus, and a set of axioms was defined which enabled scientists to derive both the previously developed laws and a large body of new knowledge.

Of course, Newton's physics only approximates observable data. Since Newton's time other theories of physics have been developed either to explain particular phenomena or to try to axiomatize all natural phenomena.

Physical theories, as well as virtually all scientific theories, especially more modern ones, are highly mathematical; and some, such as quantum theory, border on being pure mathematics. Why is this? There are several reasons.

First, the scientist whose science pertains to observable data must engage in a great deal of abstraction. Any object may have very many properties, but different scientists will be interested in different properties. For example, the chemist is interested in the chemical composition of a meteorite, the astronomer may want to know only where the meteorite originated, the physicist may be interested in the weight or density of the material. Insofar as science implies abstraction, it implies mathematics. A physical theory, like a mathematical theory, actually has existence only in the mind of the physicist, but the two kinds of theories are judged by different criteria. The physical theory must be a valid explainer and predicter of phenomena, while the mathematical theory can be valid even if it has no practical uses. Calculus will long outlive Newtonian physics.

Second, science, like any body of knowledge, needs a language. Because of its concise, logical, and abstract nature, mathematics is the perfect language for science. Thus it is not surprising that great advances in physics and other sciences have generally accompanied or followed great advances in mathematics.

Third, scientists have frequently found models in the realm of pure mathematics, which explain the phenomena they are studying. That is, the scientist interprets the axioms for a certain mathematical theory so that they form a system of axioms for the phenomenon. When he can apply the axioms to his project, then the scientist is free to use any theorems which the mathematician had derived with respect to that system of axioms.

12.3 MATHEMATICS AS AN ART

Since mathematics does not depend on its predictive value for its justification, it has a greater degree of freedom than most sciences. The mathematician is the one who creates mathematics and hence can rightly be said to be an artist.

Just as there are poor paintings and poor music, there is poor mathematics. We distinguish between invalid mathematics, which is logically unsound, and poor mathematics, which is logically correct but lacks something. What is this "something?" You might as well ask what makes good music. It is a certain elegance, brilliance, originality, call it what you will, that distinguishes good mathematics. It is very genuinely a question of esthetics.

Practical value does not make mathematics necessarily good. No one would say that a compilation of the integers from 1 to 100 is great mathematics, but the integers have great practical value.

Often the mathematician encounters a problem he feels is interesting and he wants to solve it. Perhaps he first tries to apply those tools that he already has at his disposal. If he finds that the techniques he already knows are insufficient to solve the problem, he will investigate to see whether other techniques, with which he is not familiar, have ever been applied successfully to the problem. If not, the mathematician will try to create a tool to solve the problem.

There are many unsolved mathematical problems. Those which are easiest to understand are attacked by virtually every mathematician as soon as he can understand them. Some take a great deal of training to understand, and few mathematicians are able to even attempt a solution of these problems. The mathematician solves many problems, but new ones arise from our expanded horizons and new understanding. As in the fine arts, mathematical tastes change also. The "mainstream" today may be a quiet backwater tomorrow and completely outmoded after a few years.

12.4 CONCLUSION

The author sincerely hopes that the reader has enjoyed this book. He believes that the text presents a valid picture of modern mathematics; and that the topics presented are important and will provide the best foundation in modern mathematics. The author hopes that this text has given the reader a better appreciation of mathematics and has enabled him to think a little more clearly and creatively.

Axioms for a Logic and
a Set Theory

A. AXIOMS FOR A LOGIC

We shall take as undefined concepts: statement, value, $\sim$, $\vee$, and $\wedge$. We could actually be more economical in our use of undefined terms, that is, it would be possible to axiomatize the logic developed in this text using fewer undefined terms, but there is no particular advantage to being brief.

Axiom 1. To any statement A can be assigned one and only one of the values T or F.

Axiom 2. If A and B are statements, then $\sim A$, $A \vee B$, and $A \wedge B$ are also statements; moreover, the values of $\sim A$, $A \vee B$, and $A \wedge B$ given the values of A and B are listed by Tables 4, 5 and 6 of Chapter 2.

(Observe that in Chapter 2 we tried to motivate the tables; here we merely accept the tables as part of our axioms.)

Definition 1 (*Implication*). If A and B are statements, then $A \rightarrow B$ is defined by $\sim A \vee B$.

Definition 2 (*Logical equivalence*). If A and B are statements, then $A \equiv B$ is defined by $(A \rightarrow B) \wedge (B \rightarrow A)$.

From these axioms and definitions we can develop formally the material presented in Chapters 2 and 3 of this text.

B. AXIOMS FOR A SET THEORY

It has been shown that axiomatizing set theory presents serious, indeed insurmountable, problems. Several attempts at axiomatization have been made, but not all of these axiomatizations give the same theory of sets. The first set of axioms for a theory of sets was presented by a mathematician named Zermelo in 1908. The axiomatization we present is essentially that of Zermelo and another mathematician named Fraenkel.

The undefined notions are *set* and $\in$.

Each of the axioms is given by the name by which it is usually known.

Axiom of Extensionality. If for any sets x and y, $z \in x$ if and only if $z \in y$ for any set z, then $x = y$.

Definition 1. Given two sets x and y, then $x \subset y$ if and only if $z \in x$ implies $z \in y$ for any set z.

Axiom of the Null Set. There is a set x such that for any set y we do not have $y \in x$. We denote the set x by $\varnothing$.

Axiom of Unordered Pairs. For any sets x and y, there is a set z such that $w \in z$ if and only if $w = x$ or $w = y$.

Definition 2. We let z in the Axiom of Unordered Pairs be denoted by $\{x, y\}$. If $x = y$, then $\{x, x\}$ is merely denoted by $\{x\}$. We define the *ordered pair* (x, y) to be $\big\{\{x\}, \{x, y\}\big\}$.

Definition 3. A *function* f is a set of ordered pairs such that if $(x, y) \in f$, $(x, z) \in f$, then $y = z$. The set of x such that $(x, y) \in f$ is called the *domain* of f. The set of y is the *image* of f. Denote the image of f by I_f; if u is a set such that $I_f \subset u$, then f is said to *map into* u.

(Note: What we have called the image is called the range in some texts.)

Axiom of Union. For any set x there exists a set y such that for any set z, $z \in y$ if and only if there is a set t such that $z \in t$ and $t \in x$. (The set y is the union of all sets in x.)

Definition 4. We define $z = x \cup y$ if and only if $t \in z$ if and only if $t \in x$ or $t \in y$.

Axiom of Infinity. There is a set x such that $\varnothing \in x$ and such that if any set $y \in x$, then $y \cup \{y\} \in x$. (This axiom ensures the existence of infinite sets.)

Axiom of Replacement. If x is a set and f is a function with domain x, then the image of f is a set.

(This axiom is not stated in precisely this way, but our paraphrase renders its content fairly accurately.)

Axiom of the Power Set. For any set x there is a set y such that for any set z, $z \in y$ if and only if $z \subset x$. (This axiom says that the collection of all subsets of a set forms a set.)

Definition 5. Let f be a function with domain u. We use the notation $f(x) = y$ to denote $(x, y) \in f$. We call f a *choice function* if $f(x) \in x$ for each $x \in u$.

Axiom of Choice. If x is any set, then there is a choice function f with domain u.

The Axiom of Choice is one of the most useful and most controversial axioms in set theory. It essentially says that given any family of non-empty sets, we can select an element from each of them.

Axiom of Regularity. For any set x, $x \neq \varnothing$, there is a set y such that $y \in x$ and for any set z such that $z \in x$, we do not have $z \in y$. (This axiom ensures that no set is an element of itself.)

For another presentation of a set of axioms for a theory of sets, the reader might see the Appendix to John Kelley's *General Topology*, published by Van Nostrand.

Answers to Odd-Numbered Exercises

CHAPTER 1

Section 1.3

1. **Axiom 1.** Any two vertices determine at least one pair of vertices.

   Axiom 2. Any two pairs of vertices determine at least one vertex.

   Axiom 3. There are not more than four vertices.

   Axiom 4. No pair of vertices determines every vertex.

3. None of these satisfy all of the Axioms 1 through 4.

5. If there were a concrete example which satisfied an inconsistent set of axioms, then such an example would be intrinsically self-contradictory, which is impossible. For example, it would be like finding a "square circle."

7. (a) and (d) inconsistent   (b) and (c) consistent and all independent.

Section 1.4

1. This is a discussion question with no pat answers. Some may even disagree that the mathematician's method is different from the method employed by any of these, or that the mathematician has a distinctive method. Some suggestions that the author offers, however, are given below.

Historian: studies real facts, may or may not choose to draw conclusions from the facts compiled, nonexperimental

Biologist: studies real facts, usually seeks a unified theory from his data, experimental

Philosopher: while most claim to study reality, their approach is usually more abstract than that of a natural scientist, not always strictly logical or axiomatic in method, nonexperimental

Politician: may not be as much interested in truth or logic as practical results, pragmatic

Lawyer: seeks particular end, winning case, often regardless of "reality" or logic, actions based on objective law and its interpretations, but law is not always just or logical

3. (a) *Hypothesis:* Opposite sides of a quadrilateral are equal.
   Conclusion: The quadrilateral is a parallelogram.

   (b) *Hypothesis:* Corresponding angles of two triangles are congruent.
   Conclusion: The triangles are similar.

   (c) *Hypothesis:* A quadrilateral is a parallelogram.
   Conclusion: Opposite sides are congruent.

   (d) *Hypothesis:* Two triangles are congruent.
   Conclusion: The triangles have the same area.

   (e) *Hypothesis:* An object is a square.
   Conclusion: The object is a parallelogram.

Section 1.5

1. (a), (f), and (g) are definitions.

3. (a) Let A, B, and C be triangles. Then if A is congruent to B, and B is congruent to C, it follows that A is congruent to C.

   (b) $z = 3x + 7$ (c) $8x + 7 = 4$; find x.

CHAPTER 2

Section 2.1

1. (a) "I do this" sufficient for "I am foolish," or "I am foolish" necessary for "I do this"

(b) "It does not rain" sufficient for "I will come"

(c) "It does not rain" necessary for "I will come"

(d) "I see you" sufficient for "I will shout"

(e) "The sun shines" necessary for "It is warm"

(f) "The phone rings" sufficient for "I run to answer it"

(g) "I will wear a coat" necessary and sufficient for "It snows"

(h) "I feel happy" necessary and sufficient for "I laugh"

(i) "It is a man" is sufficient for "It can read"

(j) "It is a man" is necessary and sufficient for "It is a rational animal"

Section 2.2

1. (a) Yes   (b) Yes   (c) No   (d) No   (e) No
   (f) No   (g) Yes.   (h) No   (i) Yes

3. Suppose A is true. Then B is false. But since C is a negation of B, C is true; hence if A is true, then C is true. Suppose C is true; then B is false. But then A is true. Hence if C is true, A is true. Consequently, A is true if and only if C is true.

Section 2.3

1. (a) Some man is not smart.

(b) Socrates is not a man.

(c) The sun can be shining when it is not warm.

(d) The sum of any two even integers can not be odd.

(e) All triangles are not isosceles.

(f) Somebody knows the trouble I've seen.

3. Yes. For by finding a counterexample to the negation of A, we show that the negation of A is false. But if the negation of A is false, then A must be true.

Section 2.4

1. The problem lies in the careless use of the word "each." More carefully phrased the argument would read: Any given integer is either even or odd. But not every integer is even, nor is every integer odd. We see then that all we can conclude is that there are both even and odd integers, and each integer must be either even or odd.

Section 2.5

1. (a) The wind is not blowing.

(b) The wind is blowing and the weather is warm.

(c) The wind is blowing, or the weather is warm.

(d) The wind is not blowing, and the weather is not warm.

(e) The wind is blowing or the weather is warm is not. In better English: It is not true that the wind is blowing or that the weather is warm; that is, the wind is not blowing, and the weather is not warm.

(f) It is not true that the wind is not blowing and the weather is not warm. That is, either the wind is blowing, or the weather is warm.

(g) If the wind is blowing, then the weather is warm.

(h) If the wind is not blowing, then the weather is not warm.

3. For example, the second row would be computed as follows: A is true and B is false. Since B is false, $\sim B$ is true. From Table 2.4 we read that the conjunction of two true statements is true; therefore $A \wedge \sim B$ is true.

5. (a) A is true (Table 2.4).   (b) A is false (Table 2.4).

(c) A is true (Table 2.5).

(d) A can be either true or false; the truth value is indeterminate (Table 2.7).

(e) A is false (Table 2.7).   (f) A can be either true or false (Table 2.10).

(g) A is true (Table 2.10).   (h) A is true (Tables 2.6 and 2.10).

(i) A is false (Tables 2.6 and 2.10).

7. (a) A = I am wrong, B = I will not admit it; $A \to B$

(b) A = A triangle is isosceles, B = It is not equilateral; $\sim A \to B$

(c) A = x is an integer, B = x is odd, C = x is even; $A \to (B \vee C)$

(d) A = You ran away, B = You were ignorant, C = You were foolish; $A \to (B \vee C)$

(e) A = It is not raining, B = I am seeing things, C = I am hearing things; $A \to (B \wedge C)$

(f) A = x would do such a thing, B = x is in control of his senses; $A \to \sim B$

(g) A = I watch television, B = I am nervous, C = I am bored; $A \to (B \vee C)$

(h) A = I go, B = I will do the wrong thing, C = I want to go anyway; $(A \to B) \wedge C$

Section 2.6

1. To show that $\sim(A \wedge \sim A)$ is a tautology, construct a truth table as given below. Note that $\sim(A \wedge \sim A)$ is true regardless of whether A is true or false. Any two tautologies are logically equivalent.

Table 1

A	$\sim A$	$A \wedge \sim A$	$\sim(A \wedge \sim A)$
T	F	F	T
F	T	F	T

3. The rows of Table 2.12 can be confirmed through the use of Tables 2.4 and 2.6. The rows of Table 2.13 can be confirmed through the use of Tables 2.5 and 2.6.

5. A truth table for logical equivalence is given in Table 2. Table 3 gives a truth table for $(A \to B) \wedge (B \to A)$, while Table 4 gives a truth table for $(A \wedge B) \vee (\sim A \wedge \sim B)$. Note that for corresponding truth values of A and B, $A \equiv B$, $(A \to B) \wedge (B \to A)$, and $(A \wedge B) \vee (\sim A \wedge \sim B)$ have the same truth values, hence they are logically equivalent.

Table 2

A	B	$A \equiv B$
T	T	T
T	F	F
F	T	F
F	F	T

Table 3

A	B	$A \to B$	$B \to A$	$(A \to B) \wedge (B \to A)$
T	T	T	T	T
T	F	F	T	F
F	T	T	F	F
F	F	T	T	T

Table 4

A	B	$\sim A$	$\sim B$	$A \wedge B$	$\sim A \wedge \sim B$	$(A \wedge B) \vee (\sim A \wedge \sim B)$
T	T	F	F	T	F	T
T	F	F	T	F	F	F
F	T	T	F	F	F	F
F	F	T	T	F	T	T

CHAPTER 3

Section 3.1

1. In the following the converse will be stated first, the contrapositive second.

(a) If the angles of one triangle are congruent to the angles of another, then the two triangles are similar. If the angles of one triangle are not congruent to the angles of another, then the triangles are not similar.

(b) If −1 is larger than 0, then 5 is larger than 6. If −1 is not larger than 0, then 5 is not larger than 6.

(c) When I stay inside, the weather is bad. When I do not stay inside, the weather is not bad.

(d) If I get nervous, I am taking an examination. If I do not get nervous, I am not taking an examination.

(e) If you go, I will go. If you do not go, I will not go.

(f) When the temperature is below freezing, he wears a coat. When the temperature is not below freezing, he does not wear a coat.

(g) If you are stupid, you do things the way you do. If you are not stupid, you do not act as you do.

(h) If you are well educated, you have excellent diction. If you are not well educated, you do not have excellent diction.

(i) If he is going, he is a person I have spoken to. If he is not going, I have not spoken to him.

3. (a) Equivalent   (b) Not equivalent   (c) Not equivalent
    (d) Equivalent   (e) Equivalent   (f) Equivalent

Section 3.2

1. Table 3.3 can be verified through the use of Table 2.10.

3. (a) Use Table 2.10, row 1.

(b) If $\sim B$ is T, then B is F. Apply Table 2.10, row 4, to find that A is F; hence $\sim A$ is T.

(c) Use row 1 of Table 2.4.

(d) $(A \to B) \wedge (C \to D)$ is T implies $A \to B$ is T and $C \to D$ is T. $A \to C$ is T implies A is T or C is T. If A is T, then B is T by (a); if C is T, then D is T by (a). In any case, B is T or D is T; hence $B \vee D$ is T.

Section 3.3

1. (a) *Modus ponens*, valid   (b) *Hypothetical syllogism*, valid

(c) *Fallacy of affirming the consequent*, invalid

(d) *Modus Tollens*, valid

(e) *Fallacy of affirming the consequent*, invalid

(f) *Fallacy of denying the antecedent*, invalid

(g) As in argument (19), invalid

(h) As in argument (19), invalid

(i) As in argument (19), invalid

(j) *Modus ponens*, valid

3. We present two lines of a truth table which show that $B \to A$ and $C \to A$ can both be true, while A can be either true or false.

A	B	C	$B \to A$	$C \to A$	$B \to C$
T	T	T	T	T	T
F	F	F	T	T	T

Section 3.4

1. (a) *Disjunctive syllogism*, valid   (b) *Disjunctive syllogism*, valid

(c) Invalid   (d) *Constructive dilemma*, valid   (e) Invalid

(f) *Destructive dilemma*, valid   (g) Invalid

3. A tautology is always true; its truth value is constant. If an argument needs only to have its premises true in order for its conclusion to be true, then, since any tautologies are always true, we need concern ourselves only with the truth values of those premises which are not tautologies.

Section 3.5

1. (a) Proof by contradiction, valid   (b) Proof by contradiction, valid

(c) A valid proof by disjunctive syllogism, not a proof by contradiction, since "It is raining" is not the negation of "It is cool"

(d) Valid proof by contradiction

3. (a) Valid argument form using *modus tollens* and disjunctive syllogism; would be a proof by contradiction if B were the negation of A

(b) Invalid, uses fallacy of denying antecedent

(c) Invalid, uses fallacy of affirming the consequent.

Section 3.6

1. Any triangle is either isosceles, or not. $A \lor \sim A$ (tautology)

If any triangle is not isosceles, then
there is a triangle with no two angles congruent. $\sim A \to B$

There is a triangle with no two angles congruent. B

Any triangle is not isosceles. $\sim A$

The argument is invalid. It not only uses the fallacy of affirming the consequent, but also uses "any" in a careless fashion.

3. If all men want power, they will stop at nothing. $A \to B$

All men want power. A

All men will stop at nothing. B

This is a valid *modus ponens* argument.

If all men will stop at nothing, there will be wars. $B \to C$

All men will stop at nothing. B

There will be wars. C

Another valid *modus ponens* argument.

If there will be wars, people will be killed. $C \to D$

There will be wars. C

People will be killed. D

A third valid *modus ponens* argument.

5. If I go to the party, I will see Sue. $A \to B$

If I see Sue, she will not speak to me. $B \to C$

I will not be happy if I stay home (do not go to the party). $\sim A \to \sim B$

I must be happy, or I will not be able to do my homework. $B \lor \sim D$

I will not be able to do my homework. $\sim D$

There is no way to arrive at the stated conclusion $(\sim D)$ in a valid way from the given premises.

7. If I were a fool, I would not be a senator; and
if I were a scoundrel, I would be in jail. $(A \to B) \land (C \to D)$

I am a senator, and I am not in jail. $\sim B \land \sim D$

I am neither a fool, nor a scoundrel. $\sim A \land \sim C$

Valid argument by double application of *modus tollens*.

$\sim A \land \sim C$ is equivalent to $\sim(A \lor C)$. We use this in the following.

If he is right, then I am either a fool or a scoundrel. $E \to (A \lor C)$

I am neither a fool, nor a scoundrel. $\sim(A \lor C)$

He (the opponent) is not right. $\sim E$

This is a valid *modus tollens* argument.

If my opponent is right, I should not be elected (you should not vote for me). $E \rightarrow \sim F$

My opponent is not right. $\sim E$

You should vote for me. F

This is an invalid argument (fallacy of denying the antecedent).

9. If n is odd, then n^2 is odd; that is, if $n = 2 \cdot k + 1$, then
n^2 is also of the form $2 \cdot k' + 1$. $A \rightarrow B$

n^2 is not odd. $\sim B$

n is not odd. $\sim A$

A valid *modus tollens* argument.

n is either even or odd. $A \vee C$

n is not odd. $\sim A$

n is even. C

A valid hypothetical syllogism argument. The entire argument is a proof by contradiction.

CHAPTER 4

Section 4.1

1. (a), (b), (d), (f) are sets.

3. If we consider a statement to be any declarative sentence, then the collection of true statements is not a set; for example, it would be intrinsically impossible to decide whether or not "This statement is false" is true. Generally, a statement is *defined* to be some object to which a definite truth value can be assigned. "Statement" is generally taken as a primitive, that is, undefined, notion in axiomatizations of logic.

Section 4.2

1. (a) The set of all cows, or the set of all z such that z is a cow

(b) The set of w such that w is a white house, i.e., all white houses

(c) The set of letters a, b, and c

(d) The set consisting of the letters b and c and the set which has a as its only element

(e) Elsie is a cow

(f) The set having d as its only element is not an element of the set of letters a, b, c, and d.

(g) The set having e as its only element is an element of the set which consists of the five sets having a, b, c, d, and e as their only elements, respectively.

3. (a) $(2, 3)$ (b) $\{t \mid t \text{ is an intelligent person}\}$ (c) $\{4, 6, 7\}$

(d) $\{d \mid d \text{ is a dog}\}$ (e) $\{z \mid z \text{ is both a pony and a duck}\} = \{q \mid q \neq q\}$

(f) $\{1, -1\}$

Section 4.3

1. (a) A turtle is an amphibian.   (b) Tom Smith's house is not red.

(c) Elsie and Daisy are cows.   (d) At least one whale is not a fish.

(e) Any integer greater than 5 is also greater than 3.

(f) An integer divisible by 2 is the same as an even integer.

(g) Sam, Tom, and Sally are the children of Mr. and Mrs. Brown.

3. $\{A\} \subset S$ is the same as $A \in S$. $\varnothing$ is not the same as $\{\varnothing\}$, for $\varnothing$ is by definition the set which contains no elements at all, while $\{\varnothing\}$ contains one element $\varnothing$.

5. Let $A = \{z \mid z \text{ is a lame brown dog}\}$, $B = \{s \mid s \text{ is an animal}\}$, $C = \{\text{a dog named Charlie who has red fur}\}$, $D = \{t \mid t \text{ is red}\}$, and $E = \{s \mid s \text{ is a living thing}\}$. Then we have $A \subset B$, $A \subset E$, $B \subset E$, $C \subset D$, $C \subset B$, $C \subset E$.

7. If x is an element of S, then x is an element of T. But each element of T is an element of W; hence x is an element of W. Therefore each element of S is an element of W. If x is an element of W, then x is also an element of S. We have then that $S \subset W$ and $W \subset S$, hence $S = W$.

Section 4.4

1. (a) $\{1, 2, 3, 4, a, b\} = \{a, b, 1, 2, 3, 4\}$ (b) $\{1, 2, 3, 4\} \subset \{1, 2, 3, 4, a, b\}$

(c) $\{1, 2\} \subset \{1, 2, a, b\}$ (d) Both sets are $\{1, 2, 3\}$.

(e) Both sets are $\{1, 2, 3, 4, a\}$.   (f) Both sets are $\{1, 2\}$

(g) Both sets are empty.

3. (a) x is an element of S and T if and only if x is an element of T and S.

(b) Any element of S is also an element of either S or T.

(c) If x is an element of both S and T, then x is an element of S

(d) There is no element common to both S and $\varnothing$, since $\varnothing$ contains no elements; hence $S \cap \varnothing = \varnothing$

(e) $S \cap \varnothing$ consists of all objects in either S or $\varnothing$; but since there are no objects in $\varnothing$, we have only the elements of S.

(f) $S \cap (T \cap W)$ consists of all elements in both S, and T and W, while $(S \cap T) \cap W$ consists of all elements in both S and T, and W; but these are clearly the same.

(g) $S \cup S$ consists of all elements in either S or S, that is, precisely the elements of S.

(h) $S \cap S$ consists of all elements in both S and S, that is, precisely the elements of S.

(i) If x is in $S \cup (T \cap W)$, then $x \in S$ or $x \in T \cap W$. If $x \in S$, then $x \in S \cup T$ and $x \in S \cup W$; hence $x \in (S \cup T) \cap (S \cup W)$. If $x \in T \cap W$, then $x \in T$ and $x \in W$, hence again $x \in S \cup T$ and $x \in S \cup W$; therefore $x \in (S \cup T) \cap (S \cup W)$. Thus, $S \cup (T \cap W) \subset (S \cup T) \cap (S \cup W)$. If $x \in (S \cup T) \cap (S \cup W)$, then $x \in S \cup T$, and $x \in S \cup W$. Thus, x is in S, or x is in both T and W; that is, $x \in S \cup (T \cap W)$. Hence $(S \cup T) \cap (S \cup W) \subset S \cup (T \cap W)$. Consequently, $S \cup (T \cap W) = (S \cup T) \cap (S \cup W)$.

5. (a) If $S \subset T$, then if $x \in S$, then $x \in T$. Therefore since $S \cup T$ consists of all objects in S or T, this reduces to all the objects in T. Hence $S \cup T = T$. On the other hand, if $S \cup T = T$, since $S \subset S \cup T$, we have $S \subset T$.

(b) If $S \subset T$, then since $S \cap T$ consists of all objects in both S and T and each object in S is an object in T, $S \cap T = S$. If $S \cap T = S$, then since $S \cap T \subset T$, $S \subset T$.

(c) Each element of W is an element of $T \cup W$, and each element of S is an element of $T \cup W$ (since each element of S is an element of T), but then each element of either S or W, that is, each element in $S \cup W$, is an element of $T \cup W$. Therefore $S \cup W \subset T \cup W$.

(d) An element of both S and W is an element of both T and W since such an element is an element of W and each element of S is an element of T. Thus, $S \cap W \subset T \cap W$.

Section 4.5

1. (a) $\{7, 8, 9\} \subset \{5, 6, 7, 8, 9\}$ (b) $\{7, 8, 9\} \cap \{4, 11, A\} = \varnothing$

(c) Both sets are $\{4, 11\}$.   (d) Both sets are $\{4, 6, 11, A\}$.

(e) $\{4, 11\} \neq \{4, 5, 11, A\}$ (f) $S \cap W = \{5\}$, while $S - W = \{6, 7, 8, 9\}$

3. (a) $S - S$ consists of all objects of S which are not in S; there are none, hence $S - S = \varnothing$.

(b) $S - \varnothing$ consists of all elements of S which are not in $\varnothing$; but $\varnothing$ contains no elements at all, hence $S - \varnothing = S$.

(c) If $S \subset T$, then each element of S is an element of T; but $S - T$ consists of all elements of S which are not elements of T, hence $S - T = \varnothing$. If $S - T = \varnothing$, then, since we are left with no elements at all after taking the elements of T from S, each element of S must be an element of T.

(d) $S - T = \varnothing$ implies $S \subset T$ and $T - S = \varnothing$ implies $T \subset S$ [both by (c)].

(e) If $x \in T - (S \cap W)$, then $x \in T$, but x is either not in S, or not in W. Hence x is either in $T - S$ or $T - W$; that is, $x \in (T - S) \cup (T - W)$. Thus, $T - (S \cap W) \subset (T - S) \cap (T - W)$. If $x \in (T - S) \cup (T - W)$, then $x \in T$ and either x is not in S, or x is not in W. Thus, $x \in T$, but not in $S \cap W$. Consequently, $x \in T - (S \cap W)$, and $(T - S) \cap (T - W) \subset T - (S \cap W)$.

(f) $\varnothing - S$ consists of all elements of $\varnothing$ (not in S), but $\varnothing$ contains no elements, hence $\varnothing - S = \varnothing$.

(g) $S - T$ contains no elements of T, hence $(S - T) \cap T = \emptyset$.

5. If $x \in S - (S - T)$, then $x \in S$, but $x \notin S - T$; that is, x is an element of S which is not an element of S not contained in T. Hence x is an element of S which is in T; therefore $S - (S - T) \subset T$. If $x \in T$, then $x \in S$ since $T \subset S$. But also $x \notin S - T$; hence $x \in S - (S - T)$. Therefore $T \subset S - (S - T)$.

CHAPTER 5

Section 5.1

1. (a)   (b)   (c)   (d)

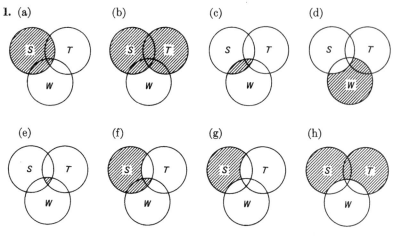

(e)   (f)   (g)   (h)

3. A formal proof of (3) is given in Section 4.5, Exercise 5. We now prove $T - (S \cup W) = (T - S) \cap (T - W)$. If $x \in T - (S \cup W)$, then $x \in T$, but x is neither in S nor W. Therefore x is in $T - S$ and $T - W$; hence $x \in (T - S) \cap (T - W)$. Therefore

$$T - (S \cup W) \subset (T - S) \cap (T - W).$$

If $x \in (T - S) \cap (T - W)$, then x is both in $T - S$ and $T - W$. Thus, $x \in T$, but x is neither in S nor W; hence $x \in T$, but $x \notin S \cup W$. Therefore, $x \in T - (S \cup W)$, and $(T - S) \cap (T - W) \subset T - (S \cup W)$.

5.

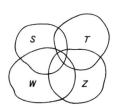

Diagram for dealing with
four arbitrary sets

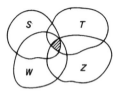

(a) $S \cap (T \cap (W \cap Z)) = (S \cap T) \cap (W \cap Z)$

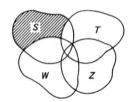

(b) $S - (T \cup (W \cup Z)) = (S - T) \cap ((S - W) \cap (S - Z))$

Section 5.2

1.

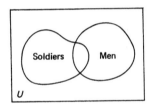

(a) $\{s \mid s$ is a soldier$\} \cap \{m \mid m$ is a man$\} \neq \emptyset$

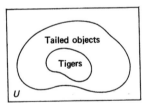

(b) $\{t \mid t$ is a tiger$\} \subset \{x \mid x$ has a tail$\}$

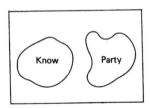

(c) $\{x \mid I$ know $x\} \cap \{y \mid y$ is going to the party$\} = \emptyset$

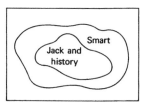

(d) $\{y \mid y$ is Jack and y takes history$\} \subset \{w \mid w$ is smart$\}$

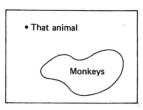

(e) That animal $\notin \{m \mid m$ is a monkey$\}$

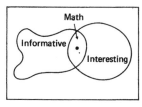

(f) Math $\in \{x \mid x$ is interesting$\} \cap \{y \mid y$ is informative$\}$

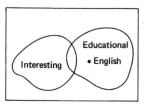

(g) English $\in \{w \mid w$ is interesting$\} \cup \{q \mid q$ is educational$\}$
(*Note:* The position of English in the diagram is not fully determined.)

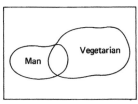

(h) $\{m \mid m$ is a man$\} \not\subset \{v \mid v$ is a vegetarian$\}$

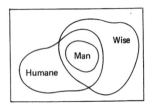

(i) $\{m \mid m \text{ is a man}\} \subset \{w \mid w \text{ is wise}\} \cap \{h \mid h \text{ is humane}\}$

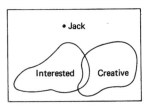

(j) Jack $\notin \{x \mid x \text{ is interested}\} \cup \{y \mid y \text{ is creative}\}$

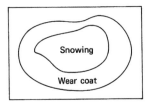

(k) $\{t \mid t \text{ is a time when I wear my coat}\} \subset \{w \mid w \text{ is a time when it is snowing}\}$

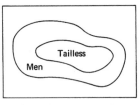

(l) $\{x \mid x \text{ is a tailless animal}\} \subset \{m \mid m \text{ is a man}\}$

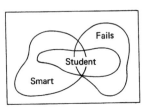

(m) $\{x \mid x \text{ is a student}\} \subset \{y \mid y \text{ is smart}\} \cup \{w \mid w \text{ fails at least one course}\}$

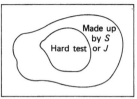

(n) $\{t \mid t$ is a hard test$\} \subset$
$\{q \mid q$ is a test made up by either Mr. Smith or Mr. Jones$\}$

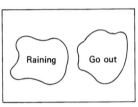

(o) $\{t \mid t$ is a time when it is raining$\} \cap \{w \mid w$ is a time when I go outside$\} = \varnothing$

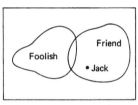

(p) Jack taking English $\in \{f \mid f$ is foolish$\} \cup \{h \mid h$ is a friend of the instructor$\}$
(The position of Jack is not fully determined.)

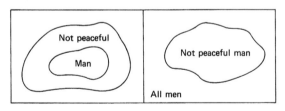

(q) $\{x \mid x$ is a man and x is not peaceful$\} \neq \varnothing$; $\{y \mid y$ is a man$\} \subset$
$\{y \mid y$ is not peaceful$\}$

(r) Sue $\in \{x \mid x$ is creative$\} \cap (\{y \mid y$ is interested$\} \cup \{w \mid w$ is hardworking$\})$
(The position of Sue is not fully determined.)

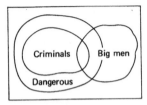

(s) $\{c \mid c$ is a criminal$\} \not\subset \{y \mid y$ is a big man$\}$; $\{c \mid c$ is a criminal$\} \subset$
$$\{d \mid d \text{ is dangerous}\}$$

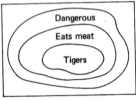

(t) $\{t \mid t$ is a tiger$\} \subset \{w \mid w$ eats meat$\}$; $\{w \mid w$ eats meat$\} \subset \{d \mid d$ is dangerous$\}$

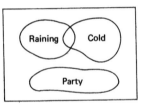

(u) $(\{t \mid t$ is a time when it is raining$\} \cup \{w \mid w$ is times when it is cold$\}) \cap$
$$\{q \mid q \text{ is a time when I go to the party}\} = \varnothing$$

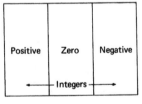

(v) $\{w \mid w$ is an integer$\} \subset \{x \mid x$ is positive$\} \cup \{0\} \cup \{z \mid z$ is negative$\}$

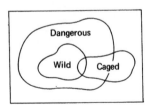

(w) $\{w \mid w$ is a wild animal$\} \subset \{d \mid d$ is dangerous$\}$; $\{w \mid w$ is a wild animal$\} \cap$
$$\{c \mid c \text{ is kept in a cage}\} \neq \varnothing$$

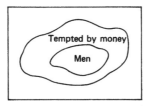

(x) $\{x \mid x \text{ is a man}\} \subset \{y \mid y \text{ can be tempted by money}\}$

3.

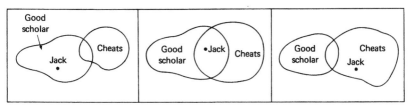

(a) $\text{Jack} \in \{x \mid x \text{ is a good scholar}\} \cup \{y \mid y \text{ cheats}\}$

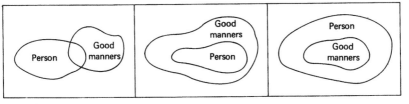

(b) $\{x \mid x \text{ is a person}\} \cap \{w \mid w \text{ has good manners}\} \neq \varnothing$

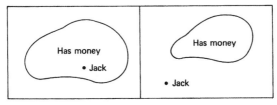

(c) $\text{Jack} \in \{x \mid x \text{ has a lot of money}\} \cup \{y \mid y \text{ does not have a lot of money}\}$

Section 5.3

1. (a) $\{x \mid x \text{ is a man}\} \subset \{y \mid y \text{ is ambitious}\}$

$\{y \mid y \text{ is ambitious}\} \subset \{w \mid w \text{ is doomed to disappointment}\}$

$\{x \mid x \text{ is a man}\} \subset \{w \mid w \text{ is doomed to disappointment}\}$

A valid argument

(b) $\{t \mid t \text{ is a tiger}\} \subset \{d \mid d \text{ is dangerous}\}$

  $\dfrac{\{d \mid d \text{ is dangerous}\} \cap \{x \mid x \text{ is in New York}\} \neq \varnothing}{\{t \mid t \text{ is a tiger}\} \cap \{x \mid x \text{ is in New York}\} \neq \varnothing}$

An invalid argument

(c) $\{x \mid x \text{ is a man}\} \subset \{h \mid h \text{ has wanted to be king}\}$

  $\dfrac{\{k \mid k \text{ is a king}\} \subset \{u \mid u \text{ is unhappy}\}}{\{x \mid x \text{ is a man}\} \cap \{p \mid p \text{ has wanted to be happy}\} \neq \varnothing;}$

An invalid argument

(d) $\{t \mid t \text{ is a tiger}\} \cap \{k \mid k \text{ is in Kansas}\} = \varnothing$

  $\dfrac{\{s \mid s \text{ is a steer}\} \cap \{k \mid k \text{ is in Kansas}\} \neq \varnothing}{\{s \mid s \text{ is a steer}\} \not\subset \{t \mid t \text{ is a tiger}\}}$

A valid argument

(e) $\{g \mid g \text{ is gold}\} \cap \{t \mid t \text{ is in Texas}\} = \varnothing$

  $\dfrac{\{r \mid r \text{ is oil}\} \subset \{t \mid t \text{ is in Texas}\}}{\{g \mid g \text{ is gold}\} \cap \{r \mid r \text{ is oil}\} = \varnothing;}$

A valid argument

(f) $\{s \mid s \text{ is Sam and robs the bank}\} \subset \{x \mid x \text{ is caught}\}$

  $\dfrac{\{x \mid x \text{ is caught}\} \subset \{k \mid k \text{ will go to jail}\}}{\{s \mid s \text{ is Sam and robs the bank}\} \subset \{k \mid k \text{ will go to jail}\};}$

A valid argument

(a)   (b)   (c)

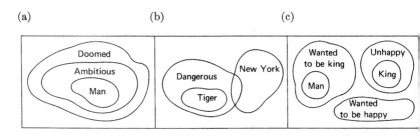

(d)   (e)   (f)

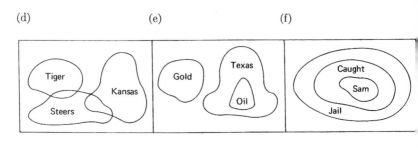

CHAPTER 6

Section 6.1

1. We give only one correct pairing; pairings other than those given may also be correct.

(a) $G \leftrightarrow K, 7 \leftrightarrow 6$ (b) $4 \leftrightarrow 0, 5 \leftrightarrow 1, 6 \leftrightarrow 2, 7 \leftrightarrow 3$

(c) Pair the ith letter of the alphabet with the number i, $i = 1, 2, \ldots, 26$

(d) Pair the positive integer n with $3n$.

(e) Pair the positive integer n with $n - 1$.

3. If S is a nonempty set, then S contains some element x. If S and $\varnothing$ had the same number of elements, then x would have to be paired with some element of $\varnothing$; but $\varnothing$ contains no elements.

Section 6.2

1. Any element s of S is paired with precisely one element t of T, but t is paired with precisely one element w of W; hence s is paired only with w. If w is any element of W, then w is paired (by π') with exactly one element t of T, but t is paired (by π) with exactly one element s of S; hence w is paired only with s. Thus, each element of S is paired with precisely one element of W, and each element of W is paired with precisely one element of S.

3. The actual pairing is described in the hint. Each subset is paired with precisely one decimal; for if W is paired with decimals D and D', then D has exactly the same digits as D', hence $D = D'$. If $.d_1d_2d_3 \ldots$ is any unending decimal whose digits are either 0 or 1, then $.d_1d_2d_3 \ldots$ has been paired with just the set W determined by the digits of $.d_1d_2d_3$; that is, 1 will be in W if and only if $d_1 = 1$, 2 will be in W if and only if $d_2 = 1$, etc.

5. Let S and T be finite sets. Then $S \cup T = S \cup (T - S)$. Now S and $T - S$ contain no elements in common, and $T - S \subset T$. Since $T - S$ is a subset of a finite set, it is finite (we will prove this next). We may assume $S = \{1, 2, 3, \ldots, n\}$ and $T - S = \{n + 1, n + 2, \ldots, n + m\}$. Then $S \cup T = \{1, 2, \ldots, n + m\}$, which is finite.

We now prove that any subset of a finite set is finite. Suppose S is a finite set. Then either $S = \varnothing$, or we may suppose $S = \{1, 2, \ldots, n\}$. If $S = \varnothing$, then the only subset of S is $\varnothing$, which is finite. Assume $S = \{1, \ldots, n\}$, and $W \subset S$. Then if $W = \varnothing$, W is finite. If $W \neq \varnothing$, then $W = \{i_1, i_2, \ldots, i_m\}$ for certain integers $i_1, \ldots, i_m$ between 1 and n. Then W has the same number of elements as $\{1, \ldots, m\}$, hence is finite.

Section 6.3

1. Suppose S contains an uncountable subset T, but S is countable. Then T is a subset of a countable set, and hence is countable by Proposition 5. But this contradicts the assumption that T is uncountable.

3. (a) *Corrected statement:* Any two *infinite* subsets of the set of positive integers contain the same number of elements. *Proof:* If W and W' are both infinite subsets of the set of positive integers, then both W and W' have the same number of elements as the set of positive integers (Proposition 3), and hence W has the same number of elements as W'.

(b) True. *Proof:* The set Z of integers is countable (corollary to Proposition 6), hence Z has the same number of elements as N, the set of positive integers. But then any infinite subset of Z is infinite and countable and hence also has the same number of elements as N.

(c) *Corrected statement:* If W is any nonempty subset of N, the set of positive integers, then ... *Proof:* If W is finite, we may assume $W = \{1, \ldots, n\}$. Then

$$W_1 = \{n + 1, \ldots, n + n = 2n\},$$
$$W_2 = \{2n + 1, \ldots, 3n\},$$
$$W_3 = \{3n + 1, \ldots, 4n\}, \text{ etc.}$$

all contain the same number of elements as W, but are distinct. There are, of course, infinitely many such sets. If W is infinite, then since all infinite subsets of N contain the same number of elements, it suffices to prove that there are infinitely many infinite subsets of N. Let $N_i = N - \{i\}$, $i = 1, 2, 3, \ldots$ Thus,

$$N_1 = N - \{1\} = \{2, 3, 4, \ldots\}, \quad N_2 = \{1, 3, 4, 5, \ldots\}, \ldots$$

Each N_i is distinct from all others and each is infinite; and, of course, there are infinitely many since there is one for each positive integer i.

(d) True. *Example.* The set of all real numbers between 0 and 1 is uncountable, but it contains the countable subset $\{1/n \mid n = 1, 2, 3, 4, \ldots\}$.

(e) *Corrected statement:* If S is a finite set, then S does not contain the same number of elements as the collection of subsets of S. *Proof:* If S contains n elements, then the collection of subsets of S contains 2^n elements. We might note that S contains fewer elements than its collection of subsets. For if $S = \{1, \ldots, n\}$, then $\{1\}, \ldots, \{n\}$ form a subset of the collection of subsets of S, but there is at least one other member of this collection, namely $\varnothing$.

Section 6.4

1. Any set is in one and only one cardinal number. The sets should be grouped as follows:

$\varnothing$ with $\{x \mid x^2 = -1,$ and x is an integer$\}$

$\{3, 4\}$ with $\{y \mid y^2 = 1,$ and y is an integer$\}$

$\{A, B, C, D, E, F\}$ with $\{h, g, j\} \cup \{k, m, n\}$

$\{z \mid z$ is a letter of the English alphabet$\}$ with

$\{R, S, T\} \cup \{x \mid x$ is an integer between 1 and 23, inclusive$\}$

N by itself

$\{z \mid z$ is less than 1, but greater than 0$\}$

(Note that the last set includes the set T of Example 4, and hence is uncountable.)

3. Let S be any set. Then S is in $\overline{S}$. Suppose S is also in $\overline{T}$. Then $\overline{S} \cap \overline{T} \neq \varnothing$, hence $\overline{S} = \overline{T}$ by the corollary to Proposition 9.

5. $\overline{\varnothing}$

Section 6.5

1. Suppose $x \in S \cup T$. Then $x \in S$, or $x \in T$, but not both, since $S \cap T = \varnothing$. If $x \in S$, then x is paired (by π_1) with precisely one element of U, and hence precisely one element of $U \cup V$. If $x \in T$, then x is paired (by π_2) with one element of V, and hence precisely one element of $U \cup V$. Therefore each element of $S \cup T$ is paired with precisely one element of $U \cup V$. Suppose $y \in U \cup V$. Then $y \in U$, or $y \in V$, but not both, since $U \cap V = \varnothing$. If $y \in U$, then y is paired with precisely one element of S; if $y \in V$, then y is paired with precisely one element of T. In either case, y is paired with precisely one element of $S \cup T$. Therefore $S \cup T$ and $U \cup V$ have the same number of elements.

3. (a) Let S be any member of $\mathfrak{C}$. Then S has the same number of elements as S.

(c) Since $\mathfrak{C} \leq \mathfrak{B}$, there are members U and V of $\mathfrak{C}$ and $\mathfrak{B}$, respectively, such that U has the same number of elements as a subset W of V. Since $\mathfrak{C} \leq \mathfrak{D}$, there are members S and T of $\mathfrak{C}$ and $\mathfrak{D}$, respectively, such that S has the same number of elements as a subset Z of T. S and T can also be chosen so that $S \cap U = \varnothing$ and $T \cap V = \varnothing$. Then $S \cup U$ is a member of $\mathfrak{C} + \mathfrak{B}$, and $T \cup V$ is a member of $\mathfrak{C} + \mathfrak{D}$. But $S \cup U$ has the same number of elements as $Z \cup W \subset T \cup V$. Therefore $\mathfrak{C} + \mathfrak{B} \leq \mathfrak{C} + \mathfrak{D}$.

(b) follows from (c) since $\mathfrak{C} \leq \mathfrak{C}$.

CHAPTER 7

Section 7.1

1. (a) $(1, a)$, $(1, b)$, $(1, c)$

(b) $(6, a)$ $\quad$ $(6, b)$ $\quad$ $(6, c)$ $\quad$ $(6, d)$ $\quad$ $(6, e)$
$\quad\,$ $(7, a)$ $\quad$ $(7, b)$ $\quad$ $(7, c)$ $\quad$ $(7, d)$ $\quad$ $(7, e)$

(c) $(-1, -1)$ $\quad$ $(-1, 0)$ $\quad$ $(-1, 1)$
$\quad\,$ $(0, -1)$ $\qquad$ $(0, 0)$ $\qquad$ $(0, 1)$
$\quad\,$ $(1, -1)$ $\qquad$ $(1, 0)$ $\qquad$ $(1, 1)$

(d) $(\%, \cup)$ $\quad$ $(\%, \cap)$ $\quad$ $(\%, *)$
$\quad\,$ $(\$, \cup)$ $\qquad$ $(\$, \cap)$ $\qquad$ $(\$, *)$
$\quad\,$ $(\#, \cup)$ $\qquad$ $(\#, \cap)$ $\qquad$ $(\#, *)$

3. $S \times \varnothing$ consists of all ordered pairs of the form (x, y) such that $x \in S$ and $y \in \varnothing$. But there are no elements in $\varnothing$, hence we cannot form any such pairs.

5. (a) Although the sets on both sides of the equality sign have very similar properties, technically we do not have equality. For example, let $S = T = W = \{1\}$. Then $(S \times T) \times W = \{((1, 1), 1)\}$, while $S \times (T \times W) = \{(1, (1, 1))\}$. The ordered pairs $((1, 1), 1)$ and $(1, (1, 1))$ are not the same.

(b) The equality is valid; we now supply a proof. Suppose $(x, w) \in (S \cup T) \times W$. Then x is an element of either S or T, and $w \in W$. Therefore (x, w) is either an element of $S \times W$, or $T \times W$; that is, $(x, w) \in (S \times W) \cup (T \times W)$. Suppose (z, w) is in $(S \times W) \cup (T \times W)$. Then $(z, w) \in S \times W$, or $(z, w) \in T \times W$. Hence $z \in S$, or $z \in T$; that is, $z \in S \cup T$. Consequently, $(z, w) \in (S \cup T) \times W$.

(c) Is also a valid equality. The proof for (b) works if $\cup$ is replaced by $\cap$, and *or* is replaced by *and*.

(d) Is valid. *Proof:* Suppose $(x, w) \in (S - T) \times W$. Then $x \in S$, but $x \notin T$, and $w \in W$. Therefore $(x, w) \in S \times W$, but $(x, w) \notin T \times W$; that is, $(x, w) \in (S \times W) - (T \times W)$. If $(x, w) \in (S \times W) - (T \times W)$, then $(x, w) \in S \times W$, but $(x, w) \notin T \times W$; hence $x \in S$, $w \in W$, but $x \notin T$. Thus, $x \in S - T$, and $(x, w) \in (S - T) \times W$.

Section 7.2

1. We may let $\{1, \ldots, m\}$ be a member of $\mathcal{B}$, and $\{m + 1, \ldots, m + n\}$ and $\{m + 1, \ldots, m + p\}$ be members of $\mathcal{A}$ and $\mathcal{C}$, respectively, where $m, n,$ and p are appropriate positive integers. Then

$$\{1, \ldots, m, m + 1, \ldots, m + n\} \qquad \text{and} \qquad \{1, \ldots, m, m + 1, \ldots, m + p\}$$

are members of $\mathcal{A} + \mathcal{B}$ and $\mathcal{C} + \mathcal{B}$, respectively. It is clear that in order for these two sets to have the same number of elements, that is, to have $\mathcal{A} + \mathcal{B} = \mathcal{C} + \mathcal{B}$, we must have $p = n$; hence $\{m + 1, \ldots, m + p\}$ has the same number of elements as $\{m + 1, \ldots, m + n\}$. Therefore $\mathcal{A} = \mathcal{C}$.

3. $\{1, \ldots, m\} \times N$ is a subset of $N \times N$; but since it is then a subset of a countable set, it is countable. Since it is infinite, it has the same number of elements as N. Therefore $\overline{m} \times \overline{N} = \overline{N}$.

5. Proposition 1 tells us that the product of any two finite sets is finite, and hence countable. If we have two countable sets S and T, and T has the same number of elements as N, the set of positive integers, then $S \times T = \varnothing$ if $S = \varnothing$; $S \times T$ has the same number of elements as $\{1, \ldots, m\} \times N$ if S contains m elements, and hence $S \times T$ has the same number of elements as N by Exercise 3; if S has the same number of elements as N, then $S \times T$ has the same number of elements as N (this can easily be shown using Example 8).

Section 7.3

1. Image of $g_3 = \{1, 3\}$; of $g_6 = \{2, 3\}$; of $g_7 = \{1, 3\}$; of $g_8 = \{2, 3\}$

3. Let $S = \{1, 2, \ldots, n\}$ and $T = \{1, 2\}$. Then a function from S to T can be formed by filling in the second coordinates of the ordered pairs $(1, -)$, $(2, -), \ldots, (n, -)$ with elements of T. This can be done in 2^n ways, thus there are 2^n functions. Looked at slightly differently, a function from S to T can be thought of as an ordered n-tuple $(a_1, \ldots, a_n)$, where the ith coordinate of the n-tuple is the element of T which appears as the second coordinate in $(i, -)$, $i = 1, \ldots, n$. In this way, the functions from S to T can be associated with the elements of $T \times T \cdots (n$ times$) \times T$. Repeated application of Proposition 1 shows that this product set contains 2^n elements. A function from T to S can be formed by filling in the second coordinates of $(1, -)$ and $(2, -)$ with elements of S. This can be done in n^2 ways (the number of elements in $S \times S$). If S contains m elements and T contains n elements, then there are n^m functions from S into T and m^n functions from T into S.

5. Functions from S into T: $\{(0, k), (1, k)\}$, $\{(0, m), (1, m)\}$, $\{(0, k), (1, m)\}$, $\{(0, m), (1, k)\}$

Functions from T into S: $\{(k, 0), (m, 0)\}$, $\{(k, 1), (m, 1)\}$, $\{(k, 0), (m, 1)\}$, $\{(k, 1), (m, 0)\}$

Section 7.4

1. $\{(1, 1), (2, 2), (3, 3)\}$, $\{(1, 1), (2, 3), (3, 2)\}$, $\{(1, 3), (2, 2), (3, 1)\}$, $\{(1, 2), (2, 1), (3, 3)\}$, $\{(1, 3), (2, 1), (3, 2)\}$, $\{(1, 2), (2, 3), (3, 1)\}$

3. (a) If $f(s)$ is an element of A, then s by definition is in $f^{-1}(A)$, and thus s could not be in $S - f^{-1}(A)$.

(b) For each $s \in S$, $f(s)$ is by definition in the image of f, hence $f^{-1}(W)$ is S.

(c) If $s \in f^{-1}(\{t\})$, then $f(s) = t$. Since this is true for every $s \in f^{-1}(\{t\})$, it follows that $f(f^{-1}(\{t\})) = t$.

(d) If $s \in f^{-1}(A)$, then $f(s) \in A$.

(e) From (d) we have $f(f^{-1}(A)) \subset A$. Suppose that $t \in A$. Since f is onto, there is $s \in S$ such that $f(s) = t$. Therefore $s \in f^{-1}(A)$ and $f(s) = t \in f(f^{-1}(A))$; hence $A \subset f(f^{-1}(A))$.

(f) If $(s, t) \in f$, then $(t, s) \in f^{-1}$; hence $(s, t) \in (f^{-1})^{-1}$. Hence $f \subset (f^{-1})^{-1}$. If $(x, y) \in (f^{-1})^{-1}$, then $(y, x) \in f^{-1}$; hence $(x, y) \in f$. Therefore $(f^{-1})^{-1} \subset f$.

5. f is one-one for any integers k and c, $k \neq 0$. For if $kz + c = kz' + c$, then $kz = kz'$, and hence $z = z'$. f will be onto if and only if $k = 1$.

Section 7.5

1. (a) $f \circ g = \{(a, a), (b, a), (c, a)\}$; $g \circ f = \{(a, a), (b, a), (c, a)\}$

(b) $f \circ g = \{(a, c), (b, a), (c, c)\}$; $g \circ f = \{(a, b), (b, b), (c, c)\}$

(c) $f \circ g = g \circ f = \{(a, c), (b, a), (c, b)\}$; $f^{-1} = \{(a, c), (b, a), (c, b)\} = g^{-1}$; $(f \circ g)^{-1} = (g \circ f)^{-1} = f^{-1} \circ g^{-1} = g^{-1} \circ f^{-1} = \{(a, b), (b, c), (c, a)\}$

(d) $f \circ g = f$; $g \circ f = g$

3. Suppose f and g are both one-one and $(g \circ f)(s) = (g \circ f)(s')$, where s and s' are in S. That is, suppose $(s, (g \circ f)(s))$ and $(s', (g \circ f)(s'))$ are both in $g \circ f$. In order to show that $g \circ f$ is one-one, it suffices to show that $s = s'$ since this will prove that no two distinct elements of $g \circ f$ can have the same second coordinate. Now $(g \circ f)(s) = g(f(s)) = (g \circ f)(s') = g(f(s'))$. Since g is one-one, $f(s) = f(s')$. But f is itself one-one, hence $s = s'$.

5. The domain of $g \circ f$ is S; its range is W. Let N be the set of positive integers with f and g functions from N into N defined by $f(n) = n^2$ and $g(n) = n^3$. Then $(g \circ f)(n) = n^6$. The image of g is the set of positive integers which are perfect cubes, while the image of $g \circ f$ is the set of positive integers which are perfect sixth powers. Image of $g \circ f = g(\text{image of } f)$.

CHAPTER 8

Section 8.1

1. The relations on S will be the subsets of

$$S \times S = \{(A, A), (A, B), (B, A), (B, B)\}.$$

These are

$$\varnothing, \{(A, A)\}, \{(A, B)\}, \{(B, A)\}, \{(B, B)\},$$
$$\{(A, A), (A, B)\}, \{(A, A), (B, A)\}, \{(A, A), (B, B)\},$$

$\{(A, B), (B, A)\}, \{(A, B), (B, B)\}, \{(B, A), (B, B)\}, \{(A, B), (B, A), (B, B)\}, \{(A, A), (B, A), (B, B)\}, \{(A, A), (A, B), (B, B)\}, \{(A, A), (A, B), (B, A)\}, S \times S.$

$\{(A, A), (B, B)\}, \{(A, A), (B, A), (B, B)\}, \{(A, A), (A, B), (B, B)\},$ and $S \times S$ contain (s, s) for each s in S. $\varnothing, \{(A, A)\}, \{(B, B)\}, \{(A, A), (A, B), (B, A)\}, \{(A, B), (B, A), (B, B)\}$ and $S \times S$ contain (s, s') whenever they contain (s', s).

3. (a) $D \subset R$ means $\{(s, s) \mid s \in S\} \subset R$, which in turn means $(s, s) \in R$, or s is R-related to itself, for each $s \in S$. If s is R-related to itself for each $s \in S$, then $(s, s) \in R$ for each $s \in S$, hence $D = (\text{df}) \{(s, s) \mid s \in S\} \subset R$.

(b) $R^{-1} = R$ if and only if (s, s') is in R if and only if (s', s) is in R, that is if and only if s R-related to s' implies s' R-related to s.

(c) $R \cap R^{-1} = \{(s, s') \mid (s, s') \text{ and } (s', s) \text{ are in } R\} = D = \{(s, s) \mid s \in S\}$ if and only if (s, s') and (s', s) in R implies $s = s'$, that is, if and only if s R-related to s', and s' R-related to s'' implies $s = s'$.

(d) $R \circ R = (\text{df}) \{(s, s'') \mid \text{there is } s' \in S \text{ such that } (s, s') \text{ and } (s', s'') \text{ are in } R\} \subset R$ if and only if given (s, s') and (s', s'') in R, (s, s'') is in R, that is, if and only if s R-related to s' and s' R-related to s'' implies s R-related to s''.

Section 8.2

1. (a) is an equivalence relation; (b) is an equivalence relation; (c) satisfies E2 and E3; (d) satisfies none of the properties of an equivalence relation; (e) satisfies E1 and E3; (f) is an equivalence relation; (g) is an equivalence relation.

3. (a) Any integer leaves the same remainder on division by 3 as itself. If n leaves the same remainder on division by 3 as m does, then m leaves the same remainder on division by 3 as n. If n leaves the same remainder on division by 3 as m does, and m leaves the same remainder on division by 3 as p, then n leaves the same remainder on division by 3 as p does.

(b) There are three equivalence classes corresponding to the three possible remainders on division by 3: 0, 1, and 2.

(c) Let $3k + 1$ and $3k' + 1$ be two integers which leave a remainder of 1 on division by 3. Then

$$(3k + 1) + (3k' + 1) = 3(k + k') + 2;$$

therefore their sum leaves a remainder of 2 on division by 3. If $3k + 2$ and $3k' + 2$ are integers which leave a remainder of 2 on division by 3, then

$$(3k + 2) + (3k' + 2) = 3(k + k' + 1) + 1,$$

a number which leaves a remainder of 1 on division by 3. Since

$$(3k + 2) + (3k' + 1) = 3(k + k' + 1),$$

we have that an integer which leaves a remainder of 2 on division by 3 plus an integer which leaves a remainder of 1 on division by 3 gives an integer which is evenly divisible by 3.

(d) $\bar{1}$ and $\bar{2}$ represent the equivalence class of integers which leave remainders of 1 and 2, respectively, on division by 3: $\bar{1} + \bar{1} = \bar{2}, \bar{1} + \bar{2} = \bar{0}$. A table for the addition of $\bar{0}, \bar{1},$ and $\bar{2}$ is given below.

+	$\bar{0}$	$\bar{1}$	$\bar{2}$
$\bar{0}$	$\bar{0}$	$\bar{1}$	$\bar{2}$
$\bar{1}$	$\bar{1}$	$\bar{2}$	$\bar{0}$
$\bar{2}$	$\bar{2}$	$\bar{0}$	$\bar{1}$

Section 8.3

1. $\bar{3} + \bar{3} = \bar{6} = \bar{4} + \bar{2}; \varnothing + \overline{21} = \overline{21} = \bar{7} + \overline{14}; \bar{9} + \bar{2} = \overline{11} = \overline{11} + \varnothing;$
$\overline{10} + \varnothing = \overline{10} = \bar{4} + \bar{6}$

3. (a) $(|\bar{m}, \bar{n}| + |\bar{p}, \bar{q}|) + |\bar{s}, \bar{t}| =$ (df) $|\bar{m} + \bar{p}, \bar{n} + \bar{q}| + |\bar{s}, \bar{t}| =$ (df)
$|(\bar{m} + \bar{p}) + \bar{s}, (\bar{n} + \bar{p}) + \bar{t}|$

Similarly,

$$|\bar{m}, \bar{n}| + (|\bar{p}, \bar{q}| + |\bar{s}, \bar{t}|) = |\bar{m} + (\bar{p} + \bar{s}), \bar{n} + (\bar{p} + \bar{t})|.$$

But

$$(\bar{m} + \bar{p}) + \bar{s} = \bar{m} + (\bar{p} + \bar{s}) \qquad \text{and} \qquad (\bar{n} + \bar{p}) + \bar{t} = \bar{n} + (\bar{p} + \bar{t})$$

(Proposition 12, Chapter 6).

(b) $|\bar{m}, \bar{n}| + |\bar{s}, \bar{t}| =$ (df) $|\bar{m} + \bar{s}, \bar{n} + \bar{t}| =$ (by hypothesis) $|\bar{m}, \bar{n}| + |\bar{p}, \bar{q}| =$ (df)
$|\bar{m} + \bar{p}, \bar{n} + \bar{q}|$

Hence

$$(\bar{m} + \bar{s}) + (\bar{n} + \bar{q}) = (\bar{m} + \bar{n}) + (\bar{s} + \bar{q})$$
$$= (\bar{n} + \bar{t}) + (\bar{m} + \bar{p}) = (\bar{m} + \bar{n}) + (\bar{p} + \bar{t}).$$

Applying Exercise 1 of Section 7.2, we have

$$\bar{s} + \bar{q} = \bar{p} + \bar{t};$$

hence

$$|\bar{s}, \bar{t}| = |\bar{p}, \bar{q}|.$$

(c) $|\bar{1}, \overline{\varnothing}| \cdot |\bar{m}, \bar{n}| =(df) |\bar{1} \times \bar{m} + \overline{\varnothing} \times \bar{n}, \overline{\varnothing} \times \bar{m} + \bar{1} \times \bar{n}|$
$$= |\bar{1} \times \bar{m} + \overline{\varnothing}, \overline{\varnothing} + \bar{1} \times \bar{n}| = |\bar{m}, \bar{n}|$$

(d) $|\overline{\varnothing}, \bar{1}| \cdot |\bar{m}, \bar{n}| =(df) |\overline{\varnothing} \times \bar{m} + \bar{1} \times \bar{n}, \bar{1} \times \bar{m} + \overline{\varnothing} \times \bar{n}| = |\bar{n}, \bar{m}|$

(e) $|\overline{\varnothing}, \bar{m}| \cdot |\overline{\varnothing}, \bar{n}| =(df) |\overline{\varnothing} \times \overline{\varnothing} + \bar{m} \times \bar{n}, \overline{\varnothing} \times \bar{n} + \bar{m} \times \overline{\varnothing}| = |\bar{m} \times \bar{n}, \overline{\varnothing}|$

5. $\{1, \ldots, m\}$ is a member of $\bar{m}$ and $\{1, \ldots, n\}$ is a member of $\bar{n}$.

Case 1: $\{1, \ldots, m\} = \{1, \ldots, n\}$. Then $\bar{n} = \bar{m}$ and $(\bar{m}, \bar{n})$ is equivalent to $(\overline{\varnothing}, \overline{\varnothing})$.

Case 2: $\{1, \ldots, m\}$ is contained in, but not equal to, $\{1, \ldots, n\}$. Then there is a positive integer k such that $\{1, \ldots, m, m + 1, \ldots, m + k\} = \{1, \ldots, n\}$. Then $(\bar{m}, \bar{n})$ is equivalent to $(\overline{\varnothing}, \bar{k})$.

Case 3: $\{1, \ldots, n\}$ is contained in, but not equal to, $\{1, \ldots, m\}$. Then there is a positive integer k such that

$$\{1, \ldots, n, n + 1, \ldots, n + k\} = \{1, \ldots, m\}.$$

Then $(\bar{m}, \bar{n})$ is equivalent to $(\bar{k}, \overline{\varnothing})$.

Section 8.4

1. (a) **P1.** For any integer n, $n = n$; hence (by definition) nRn.

P2. If nRm and mRn, then n is less than or equal to m in the usual sense and m is less than or equal to n. Since "less than or equal to" in its usual sense gives a partial ordering, $m = n$.

P3. This also follows from the observation that if we have mRn, then m is less than or equal to n (for m is either less than $m - 1$, and hence is less than n, or $m = n$) and the fact that "less than or equal to" defines a partial ordering.

(b) $3 \neq 4$, hence if $3R4$, or $4R3$, then 3 is strictly less than $4 - 1 = 3$, or 4 is strictly less than 2, neither of which is the case. Therefore R is not a total ordering.

(c) If there is a least upper bound for $\{3, 4\}$, then it would have to be 6. But 7 is also an upper bound for $\{3, 4\}$, and 6 is not R-related to 7. For $6 \neq 7$, neither is 6 strictly less than $7 - 1 = 6$.

(d) Given any two even integers k and k', either $k = k'$, in which case kRk', or $k \neq k'$. If $k \neq k'$, we may assume k is greater than k'. Then k is at least 2 greater than k', hence k' is strictly less than $k - 1$; hence $k'Rk$.

3. Since v is a greatest lower bound for W and v' is a lower bound, we have $v'Rv$. But since v' is a greatest lower bound and v is a lower bound, we also have vRv'; hence by P2, we have $v = v'$.

5. (a) **P1.** If $(s, t) \in S \times T$, then $(s, t)R''(s, t)$ since sRs and $tR't$.

P2. If $(s, t)R''(s', t')$ and $(s', t')R''(s, t)$, then sRs' and $s'Rs$ as well as $tR't'$ and $t'R't$. Therefore $s = s'$ and $t = t'$; hence $(s, t) = (s', t')$.

P3. If $(s, t)R''(s', t')$ and $(s', t')R''(s'', t'')$, then sRs' and $s'Rs''$ as well as $tR't'$ and $t'R't''$. Hence sRs'' and $tR't''$; consequently, $(s, t)R''(s'', t'')$.

(b) Suppose $(w, z) \in W \times Z$. Then wRu and $zR'u'$ since u and u' are upper bounds for W and Z, respectively. Therefore (u, u') is an upper bound for $W \times Z$. Suppose (v, v') is also an upper bound for $W \times Z$. Then for any $(w, z) \in W \times Z$, wRv and $zR'v'$. This in turn implies that for any $w \in W$, we have wRv and for any $z \in Z$, we have zRv'. Therefore v is an upper bound for W and v' is an upper bound for Z. Since u and u' are least upper bounds, uRv and $u'R'v'$. Thus, $(u, u')R''(v, v')$. It follows then that (u, u') is the least upper bound for $W \times Z$.

CHAPTER 9

Section 9.1

1. P1. $(x, m)R(x, m)$ since the first coordinates are the same and m is less than or equal to m.

P2. If $(x, m)R(y, n)$ and $(y, n)R(x, m)$, then x must be the same as y or one of these relationships could not hold. Moreover, we also have m less than or equal to n and n less than or equal to m; hence $m = n$. Thus $x = y$ and $m = n$, and $(x, m) = (y, n)$.

P3. If $(x, m)R(y, n)$ and $(y, n)R(z, p)$, then x is less than or equal to y and y is less than or equal to z; moreover, m is less than or equal to n, and n is less than or equal to p. Therefore x is less than or equal to z and m is less than or equal to p; hence $(x, m)R(z, p)$. We now prove that the ordering is total: Suppose (x, m) and (y, n) are in $M \times N$. Then either $x = 0$ and $y = 1$, in which case $(x, m)R(y, n)$ regardless of what m and n are; or $x = 1$ and $y = 0$, in which case $(y, n)R(x, m)$; or $x = y$. If $x = y$, and m is less than or equal to n, then $(x, m)R(y, n)$; if $x = y$, and n is less than or equal to m, then $(y, n)R(x, m)$.

3. (a) In order to have $|\overline{\varnothing}, \overline{\varnothing}| < |\overline{m}, \overline{n}| < |\overline{1}, \overline{\varnothing}|$, we would have to have $\overline{n} + \overline{\varnothing} < \overline{m} + \overline{\varnothing}$, and hence $\overline{n} < \overline{m}$, or $\{1, \ldots, n\}$ properly contained in $\{1, \ldots, m\}$, as well as $\overline{m} + \overline{\varnothing} = \overline{m} \leq \overline{n} + \overline{1}$, or $\{1, \ldots, m\}$ properly contained in $\{1, \ldots, n + 1\}$. We thus have $\{1, \ldots, n + 1\} \subset \{1, \ldots, m\}$ since $\{1, \ldots, n\}$ is a proper subset of $\{1, \ldots, m\}$ and $\{1, \ldots, m\}$ is a proper subset of $\{1, \ldots, n + 1\}$. Hence $\{1, \ldots, m\}$ is a proper subset of itself, a contradiction.

(b) Since $\{1, 2, \ldots, m + n\}$ is a proper subset of $\{1, 2, \ldots, m + n + 1\}$, we have $\overline{m} + \overline{n} < \overline{m} + \overline{n} + \overline{1}$; hence $|\overline{m}, \overline{n}| < |\overline{m} + \overline{1}, \overline{n}|$. If there were an integer $|\overline{p}, \overline{q}|$ between $|\overline{m}, \overline{n}|$ and $|\overline{m} + \overline{1}, \overline{n}|$, then there would have to be a set between $\{1, \ldots, m + n\}$ and $\{1, \ldots, m + n + 1\}$, which is impossible. A formal argument to this effect can be had from (a) after making the appropriate changes. The immediate predecessor of $|\overline{m}, \overline{n}|$ is $|\overline{m}, \overline{n} + \overline{1}|$.

5. (a) For $n = 1$, $1 < 2$ (which we will assume as being true, but which we could prove using the definition on the integers given in Chapter 8 and the total ordering for the integers given in Exercise 6 of Section 8.3). Assume that $n < n + 1$ for some positive integer n. Adding 1 to both sides of the inequality, we have $n + 1 < n + 2$; that is, $n + 1 < (n + 1) + 1$. Thus, if the inequality applies to n, it also applies to $n + 1$. Consequently, the inequality is true for any positive integer.

(b) For $n = 1$, $1^2 + 1 = 2$, which is divisible by 2. Assume $n^2 + n$ is divisible by 2. Now

$$(n + 1)^2 + (n + 1) = n^2 + 2n + 1 + n + 1 = (n^2 + n) + 2(n + 1).$$

Since $n^2 + n$ is divisible by 2 by hypothesis and $2(n + 1)$ is divisible by 2, $(n + 1)^2 + (n + 1)$ is divisible by 2. Therefore for any positive integer n, $n^2 + n$ is divisible by 2.

7. If n is a lower bound for a subset W of integers and m is an upper bound for W, then $W \subset \{n, n + 1, n + 2, \ldots, n + t = m\}$, a finite set (since m is reached after only t steps). Therefore since W is a subset of a finite set, it is finite. This property is not true of the real numbers; for example, there are infinitely many real numbers between 0 and 1 (see Example 4, Chapter 6).

Section 9.2

1. Suppose t and t' are both immediate predecessors of s, but $t \neq t'$. Then t is less than t', or t' is less than t, and both t and t' are less than s. If t is less than t', then there is a predecessor of s which is greater than t, hence t is not an immediate predecessor. If t' is less than t, then there is a predecessor of s which is greater than t', hence t' is not an immediate predecessor. It must be then that $t = t'$.

3. Suppose s is the first element of D, a discrete set, and t is the last element. Since t can be reached from s by taking immediate successors in only finitely many steps, we have a subset

$$\{s_0 = s, s_1 = s^+, \ldots, s_{n-1}, s_n = (s_{n-1})^+ = t\}$$

of D, and this subset is finite. But this subset contains every element of D. For if w is any element of D, then we can reach w from s by taking immediate successors in finitely many steps. But these immediate successors are unique (Proposition 2), hence w is s_i for some i, $i = 0, \ldots, n$. Consequently, D contains only finitely many elements.

Suppose now that D is a discrete set with finitely many elements. Choose any w in D. If w has no immediate successors, then w is a last element of D. If w has successors, we proceed by immediate successors from w until we reach an element t of D which has no successors. There must be such an element or D would be infinite; t is a last element. Working from w and taking immediate predecessors, we eventually reach a first element of D.

5. People waiting in line; a number of procedures to be carried out in a definite order; the hierarchical structure of a business organization provided no two people have the same rank; etc.

Section 9.3

1. (a) The progression we will use is the set of positive integers. For $n = 1$, $1 < 2 \cdot 1 = 2$; hence the statement is true for $n = 1$.* Assume $n < 2n$ for some positive integer n. Adding 1 to both sides of the inequality, we obtain $n + 1 < 2n + 1$. And $2n + 1 < (2n + 1) + 1 = 2n + 2 = 2(n + 1)$. Therefore $n + 1 < 2n + 1 < 2(n + 1)$. Since the statement is true for $n = 1$ and is true for $n + 1$ whenever it is true for n, it is true for each positive integer.

(b) The progression used is the set of positive integers. For $n = 1$, $4 > 3$. Assume that for some positive integer n, $4n > 2n + 1$. Adding 4 to both sides, we get $4n + 4 = 4(n + 1) > 2n + 1 + 4 = 2(n + 1) + 3 > 2(n + 1) + 1$.

(c) The progression used is the set of positive integers. For $n = 1$,

$$2 \text{ (the sum of the first 1 positive even integers)} = 1 \cdot (1 + 1) = 2.$$

Assume that the sum of the first n positive even integers is $n(n + 1)$. Then the sum of the first $n + 1$ positive even integers is

$$2 + 4 + \cdots + 2n + 2(n + 1),$$

which is the sum of the first n positive even integers with $2(n + 1)$. Since we have assumed that the sum of the first n positive even integers is $n(n + 1)$, then the sum of the first $n + 1$ positive even integers is

$$n(n + 1) + 2(n + 1) = (n + 2)(n + 1) = (n + 1)((n + 1) + 1),$$

which is the formula applied to $n + 1$.

(d) The progression used is the set of positive integers. For $n = 1$, the sum of the first 1 positive odd integer is $1 = 1^2$. Assume that the sum of the first n positive odd integers is n^2. The sum of the first $n + 1$ positive odd integers is

$$1 + 3 + \cdots + (2n - 1) + (2n + 1) = (1 + 3 + \cdots + (2n - 1)) + (2n + 1)$$
$$= n^2 + 2n + 1 = (n + 1)^2.$$

(e) The progression used is the positive integers. For $n = 1$, $1 < 1^2 + 1 = 2$. Assume that for some positive integer n, $n < n^2 + 1$. Adding 1 to both sides, we obtain $n + 1 < n^2 + 1 + 1$. But

$$n^2 + 2 < n^2 + 2 + (2n - 1) = n^2 + 2n + 1 = (n + 1)^2.$$

Therefore $n + 1 < (n + 1)^2 < (n + 1)^2 + 1$.

(f) The progression used is the set of positive integers. The sum of the first 1 perfect square is

$$1^2 = 1 = \tfrac{1}{6} \cdot 1 \cdot (1 + 1) \cdot (2 + 1).$$

Assume that for some positive integer n,

$$1^2 + 2^2 + \cdots + n^2 = \tfrac{1}{6}n(n + 1)(2n + 1).$$

---

* $<$ stands for "is less than," $>$ stands for "is greater than."

Then

$$(1^2 + 2^2 + \cdots + n^2) + (n+1)^2 = \tfrac{1}{6}n(n+1)(2n+1) + (n+1)^2$$
$$= \tfrac{1}{6}\big(n(n+1)(2n+1) + 6(n+1)^2\big)$$
$$= \tfrac{1}{6}(n+1)\big(n(2n+1) + 6(n+1)\big)$$
$$= \tfrac{1}{6}(n+1)(n+2)\big(2(n+1)+1\big)$$

(the fastest way to confirm this last equality is to expand this last term and the term before it); but this is the formula applied to $n+1$.

(g) The progression we use is the set of positive integers larger than 4, that is, $\{5, 6, 7, \ldots\}$. For $n = 5$, $25 < 32$. Suppose for some positive integer n, $n^2 < 2^n$. Multiplying both sides of the inequality by 2, we obtain $2n^2 < 2^{n+1}$. We now show that $(n+1)^2 < 2n^2$ if n is larger than 4. This in turn will give us

$$(n+1)^2 < 2n^2 < 2^{n+1}, \qquad \text{or} \qquad (n+1)^2 < 2^{n+1},$$

which is the desired inequality for $n+1$. For $n = 5$, $(n+1)^2 = 36 < 2n^2 = 50$. Expanding both sides of the (second) inequality and simplifying, we find that what we wish to prove reduces to $n^2 > 2n + 1$. Assume then that for a positive integer n, $n^2 > 2n + 1$. Adding $2n + 1$ to both sides and factoring, we have

$$(n+1)^2 > 4n + 2 = 2(n+1) + 2n > 2(n+1) + 1.$$

Hence the second inequality is also proved.

(h) The progression we use is $\{3, 4, 5, \ldots\}$. For $n = 3$, 3 divides $27 - 3 = 24$. Suppose 3 divides $n^3 - n$ for some positive integer n. Now

$$(n+1)^3 - (n+1) = n^3 + 3n^2 + 3n + 1 - (n+1)$$
$$= n^3 + 3n^2 + 2n = (n^3 - n) + 3(n^2 + n).$$

Since both terms of this latter sum are divisible by 3, the whole is divisible by 3; thus the truth of the statement for n implies the truth for $n+1$.

3. The sum of two positive integers is a positive integer. Suppose the sum of m positive integers is again a positive integer for $m = 2, 3, \ldots, n-1$. We now show that this implies that the sum of n positive integers is again a positive integer. Let $a_1 + a_2 + \cdots + a_n$ be the sum of n positive integers. Then

$$a_1 + a_2 + \cdots + a_n = (a_1 + a_2) + (a_3 + \cdots + a_n).$$

Each of the terms in parentheses on the right side of this equality is the sum of less than n positive integers and is thus a positive integer. But this means that $a_1 + a_2 + \cdots + a_n$ is the sum of two positive integers and is therefore a positive integer.

5. (a) The proof here is essentially the same as for 3, but with $+$ replaced by $\cdot$ and "sum" replaced by "product."

(b) If $m = 1$, then $(a \cdot b)^1 = a^1 \cdot b^1$. Suppose $(a \cdot b)^m = a^m \cdot b^m$ for some positive integer m. Then

$$(a \cdot b)^{m+1} = (a \cdot b)^m (a \cdot b)^1 = (a^m \cdot b^m) \cdot (a^1 \cdot b^1)$$
$$= (a^m \cdot a^1) \cdot (b^m \cdot b^1) = a^{m+1} \cdot b^{m+1}.$$

Section 9.4

1. (a) $s_1 = 2, s_2 = 3, s_3 = 4, s_4 = 5, s_5 = 6$

(b) $s_1 = 0, s_2 = 4, s_3 = 8, s_4 = 12, s_5 = 16$

(c) $s_1 = 2, s_2 = 4, s_3 = 16, s_4 = 256, s_5 = (256)^2$

CHAPTER 10

Section 10.1

1. (a) Not high winds   (b) Lightning and hail

(c) Hail or temperatures over 70°

(d) Lightning if the temperature is over 70°

(e) Not high winds if it is hailing   (f) Neither lightning, nor high winds

(g) Hail if there is either lightning or high winds

(h) Either lightning and high winds, or hail and temperature over 70°

(i) Not lightning, together with either hail or not high winds

(j) Lightning if there is hail, together with either high winds or a temperature over 70°

(k) Lightning, hail, and high winds

(l) Either lightning, or hail, or temperature over 70°

(m) Not lightning, and hail, and temperature over 70°

(n) Either lightning or hail if there are high winds and temperature over 70°

3. (a) 20 times   (b) 9 times   (c) 91 times

(d) $R(A) = 20/100, R(B) = 9/100, R(\overline{B}) = 91/100$

(e) $R(A, B) = 7/100$

(f) Lightning without hail occurs more than 6 times as often as hail without lightning.

(g) No

Section 10.2

1. (a) $P(A, B)$ lies between $\frac{2}{5}$ and $\frac{3}{5}$.

(b) $P(B \mid A) = \frac{1}{2}$

(c) $P(B)$ lies between $\frac{1}{4}$ and $\frac{3}{4}$.

(d) $P(\overline{A}, B) = 0.18$ (e) $P(A)$ lies between 0.4 and $\frac{1}{2}$.

3. If $P(A) = P(A \mid B)$, then the chance of A occurring is unaffected by the occurrence or nonoccurrence of B; thus, whether or not A will occur is independent of whether or not B will occur. If A and B are independent, then $P(A, B) = P(A)P(B)$.

5. (a) $P(R + L) = 0.12 = 0.1 + 0.05 - P(R, L)$; hence $P(R, L) = 0.03$.

(b) $P(R, L) = 0.03 = 0.05(P(R \mid L))$; hence $P(R \mid L) = 0.6$.

(c) $P(R, L) = 0.03 = 0.1(P(L \mid R))$; hence $P(L \mid R) = 0.3$; L and R are neither mutually exclusive nor independent.

(e) $P(\overline{R}, \overline{L}) = 1 - P(R + L) = 0.88$

(f) (i) $P(R, \overline{L}) + P(\overline{R}, L) = P(R)P(\overline{L} \mid R) + P(L)P(\overline{R} \mid L) = (0.1)(0.4) + (0.05)(0.7) = 0.075$ (ii) $P(L, \overline{R}) = P(L)P(\overline{R} \mid L) = 0.035$

(iii) $P(R + \overline{L}) = P(R) + P(\overline{L}) - P(R, \overline{L}) = 0.1 + 0.05 - 0.04 = 0.11$

Section 10.3

1. (a) $\frac{1}{36}$ (b) $\frac{3}{36}$ (c) $\frac{5}{36}$ (d) $\frac{4}{36}$ (e) $\frac{2}{36}$

3. Let $2x$ be the probability of rolling a 3 with the unfair die. Then the probability of rolling any face other than a 3 is x. Then $2x + 5x = 7x = 1$; hence $x = \frac{1}{7}$.

$$P(3 \text{ with the unfair die}) = \tfrac{2}{7}.$$

$P(7$ with two dice, one fair and the other the unfair die$)$
$$= P(2, 5) + P(5, 2) + P(6, 1) + P(1, 6) + P(3, 4) + P(4, 3)$$
$$= \tfrac{1}{42} + \tfrac{1}{42} + \tfrac{1}{42} + \tfrac{1}{42} + \tfrac{2}{42} + \tfrac{1}{42} = \tfrac{7}{42},$$

where the first coordinate in each of the pairs represents what is rolled on the unfair die.

Section 10.4

1. (a) 24   (b) 4   (c) 45   (d) 156   (e) 362,880   (f) 1

3. (a) $C(8, 4) = 70$ (b) 4^8 (c) $_8P_4 = 1680$ (d) 6^6
(e) 6^6 (f) $C(10, 5) = 216$ (g) $_{10}P_5 = 38{,}880$ (h) 5^{10}

Section 10.5

1. $\dfrac{C(8, 2)C(10, 5)}{C(18, 7)}$ **3.** $\dfrac{C(7, 1)C(5, 2)}{C(12, 3)}$ **5.** $C(12, 5)(\tfrac{1}{4})^5(\tfrac{3}{4})^7$

7. $\dfrac{(0.3)(0.3)^3}{((0.3)(0.3)^3 + (0.6)(0.6)^3 + (0.1)(0.1)^3)}$

9. $C(8000; 0)(0.9999)^{8000} + C(8000; 1)(0.0001)(0.9999)^{7999} +$
$C(8000; 2)(0.0001)^2(0.9999)^{7998} + C(8000; 3)(0.0001)^3(0.9999)^{7997}$

11. $(\tfrac{1}{4})(1)/((\tfrac{1}{4})(1) + 3(\tfrac{3}{4})(\tfrac{1}{2})^5)$

CHAPTER 11

Section 11.1

1. *Example* 2. We have defined things so that any two points determine a unique line. Any line other than L is of the form $\{P, X\}$, where X is some point of L. L intersects any line $\{P, X\}$ other than itself in X alone. Given two lines

$\{P, X\}$ and $\{P, X'\}$, $X \neq X'$, other than L, these intersect only in P. If X and X' are two distinct points of L, then X, X', and P are three distinct points not all on the same line.

Example 3. The only road open to the reader with this example is that of direct verification, that is, testing pairs of points and pairs of lines.

3. If we let $\{1, 2, 3\}$ be L and $4 = P$, we can, by imitating Example 2, arrive at a system which satisfies Axioms 1 through 3. If we try to make subsets of $\{1, 2, 3, 4\}$ into lines such that each line contains at least three points, then, since any two subsets of $\{1, 2, 3, 4\}$ which each contain three elements must overlap in two elements, points must be sets of two elements each. But this, in turn, implies that some one line contains all the elements of $\{1, 2, 3, 4\}$, or that there are not three points not all lying on the same line.

Section 11.2

1. If Y and Y' are points of L_1 paired with X of L_2, then QY and QY' both lie on X. But then QY and QY' lie on both Q and X, hence $QY = QY' = QX$. Since QX and L_1 lie on precisely one point, we must have $Y = Y'$. If Y and Y' are both points of L_2 which are paired with some point X of L_1, then QX and L_2 lie on both Y and Y'. But since L_2 and QX can lie on but a single point, we must have $Y = Y'$.

3. The "dual" of the given statement is:

Given a line L which does not lie on a point P, there is precisely one point which lies on L and is not P.

Since L can contain infinitely many points, this statement is false.

5. Example 4 shows that each line in a projective plane can lie on infinitely many points. There can be no projective plane in which each line lies on 9 points since there is no positive integer n for which $9 = n^2 - n + 1$. If each line lies on $n + 1$ points, then the plane has $(n + 1)^2 - (n + 1) + 1 = n^2 + n + 1$ points.

Section 11.3

1. *Axiom D.* There are three lines in π, each lying on at least three points. Hence the removal of one line with all its points from π still leaves us with at least two lines each lying on at least two points. Therefore we can choose two points from one line L and one point from another line L' which does not lie on L.

Axiom E follows immediately by definition of π' and Axiom 5 of π.

3. Any line L is parallel to itself. If L is parallel to L', then either $L = L'$, or L and L' lie on no common point. In either case, L' is also parallel to L. Suppose now that L is parallel to L', and L' is parallel to L''. If $L = L''$, then L is parallel to L''. If $L \neq L''$, but L and L'' lie on some point P, then L and L'' both lie on P and are parallel to L', contradicting Axiom C. Therefore if $L \neq L''$, then L and L'' lie on no common point; hence L is parallel to L''.

5. Let P be any point and L be any line which does not lie on P. Then there is one line through P for each of the n points of L as well as one line through P which is parallel to L; hence there are $n + 1$ lines lying on P.

Second method: Let π be the projective plane formed from π', an affine plane in which each line lies on n points, in accordance with Proposition 14. Then each line in π lies on $n + 1$ points, hence each point lies on $n + 1$ lines. But any point of π' only lies on lines formed from lines of π'. Hence each point of π' lies on $n + 1$ points both in π and in π'.

7. If we could be sure there was an affine plane in which each line lies on n points, then we could also be sure that there was a projective plane in which each line lies on $n + 1$ points, where n is any positive integer. But this latter information is unknown in the case of many positive integers.

Index of Symbols

Index

A CATALOG OF SELECTED
DOVER BOOKS
IN SCIENCE AND MATHEMATICS

Astronomy

BURNHAM'S CELESTIAL HANDBOOK, Robert Burnham, Jr. Thorough guide to the stars beyond our solar system. Exhaustive treatment. Alphabetical by constellation: Andromeda to Cetus in Vol. 1; Chamaeleon to Orion in Vol. 2; and Pavo to Vulpecula in Vol. 3. Hundreds of illustrations. Index in Vol. 3. 2,000pp. 6⅛ x 9¼.
Vol. I: 23567-X
Vol. II: 23568-8
Vol. III: 23673-0

EXPLORING THE MOON THROUGH BINOCULARS AND SMALL TELESCOPES, Ernest H. Cherrington, Jr. Informative, profusely illustrated guide to locating and identifying craters, rills, seas, mountains, other lunar features. Newly revised and updated with special section of new photos. Over 100 photos and diagrams. 240pp. 8¼ x 11. 24491-1

THE EXTRATERRESTRIAL LIFE DEBATE, 1750–1900, Michael J. Crowe. First detailed, scholarly study in English of the many ideas that developed from 1750 to 1900 regarding the existence of intelligent extraterrestrial life. Examines ideas of Kant, Herschel, Voltaire, Percival Lowell, many other scientists and thinkers. 16 illustrations. 704pp. 5⅜ x 8½. 40675-X

THEORIES OF THE WORLD FROM ANTIQUITY TO THE COPERNICAN REVOLUTION, Michael J. Crowe. Newly revised edition of an accessible, enlightening book recreates the change from an earth-centered to a sun-centered conception of the solar system. 242pp. 5⅜ x 8½. 41444-2

A HISTORY OF ASTRONOMY, A. Pannekoek. Well-balanced, carefully reasoned study covers such topics as Ptolemaic theory, work of Copernicus, Kepler, Newton, Eddington's work on stars, much more. Illustrated. References. 521pp. 5⅜ x 8½.
65994-1

A COMPLETE MANUAL OF AMATEUR ASTRONOMY: Tools and Techniques for Astronomical Observations, P. Clay Sherrod with Thomas L. Koed. Concise, highly readable book discusses: selecting, setting up and maintaining a telescope; amateur studies of the sun; lunar topography and occultations; observations of Mars, Jupiter, Saturn, the minor planets and the stars; an introduction to photoelectric photometry; more. 1981 ed. 124 figures. 26 halftones. 37 tables. 335pp. 6½ x 9¼.
42820-6

AMATEUR ASTRONOMER'S HANDBOOK, J. B. Sidgwick. Timeless, comprehensive coverage of telescopes, mirrors, lenses, mountings, telescope drives, micrometers, spectroscopes, more. 189 illustrations. 576pp. 5⅝ x 8¼. (Available in U.S. only.)
24034-7

STARS AND RELATIVITY, Ya. B. Zel'dovich and I. D. Novikov. Vol. 1 of *Relativistic Astrophysics* by famed Russian scientists. General relativity, properties of matter under astrophysical conditions, stars, and stellar systems. Deep physical insights, clear presentation. 1971 edition. References. 544pp. 5⅝ x 8¼. 69424-0

Chemistry

THE SCEPTICAL CHYMIST: The Classic 1661 Text, Robert Boyle. Boyle defines the term "element," asserting that all natural phenomena can be explained by the motion and organization of primary particles. 1911 ed. viii+232pp. 5⅜ x 8½.
42825-7

RADIOACTIVE SUBSTANCES, Marie Curie. Here is the celebrated scientist's doctoral thesis, the prelude to her receipt of the 1903 Nobel Prize. Curie discusses establishing atomic character of radioactivity found in compounds of uranium and thorium; extraction from pitchblende of polonium and radium; isolation of pure radium chloride; determination of atomic weight of radium; plus electric, photographic, luminous, heat, color effects of radioactivity. ii+94pp. 5⅜ x 8½. 42550-9

CHEMICAL MAGIC, Leonard A. Ford. Second Edition, Revised by E. Winston Grundmeier. Over 100 unusual stunts demonstrating cold fire, dust explosions, much more. Text explains scientific principles and stresses safety precautions. 128pp. 5⅜ x 8½.
67628-5

THE DEVELOPMENT OF MODERN CHEMISTRY, Aaron J. Ihde. Authoritative history of chemistry from ancient Greek theory to 20th-century innovation. Covers major chemists and their discoveries. 209 illustrations. 14 tables. Bibliographies. Indices. Appendices. 851pp. 5⅜ x 8½.
64235-6

CATALYSIS IN CHEMISTRY AND ENZYMOLOGY, William P. Jencks. Exceptionally clear coverage of mechanisms for catalysis, forces in aqueous solution, carbonyl- and acyl-group reactions, practical kinetics, more. 864pp. 5⅜ x 8½.
65460-5

ELEMENTS OF CHEMISTRY, Antoine Lavoisier. Monumental classic by founder of modern chemistry in remarkable reprint of rare 1790 Kerr translation. A must for every student of chemistry or the history of science. 539pp. 5⅜ x 8½. 64624-6

THE HISTORICAL BACKGROUND OF CHEMISTRY, Henry M. Leicester. Evolution of ideas, not individual biography. Concentrates on formulation of a coherent set of chemical laws. 260pp. 5⅜ x 8½.
61053-5

A SHORT HISTORY OF CHEMISTRY, J. R. Partington. Classic exposition explores origins of chemistry, alchemy, early medical chemistry, nature of atmosphere, theory of valency, laws and structure of atomic theory, much more. 428pp. 5⅜ x 8½. (Available in U.S. only.)
65977-1

GENERAL CHEMISTRY, Linus Pauling. Revised 3rd edition of classic first-year text by Nobel laureate. Atomic and molecular structure, quantum mechanics, statistical mechanics, thermodynamics correlated with descriptive chemistry. Problems. 992pp. 5⅜ x 8½.
65622-5

FROM ALCHEMY TO CHEMISTRY, John Read. Broad, humanistic treatment focuses on great figures of chemistry and ideas that revolutionized the science. 50 illustrations. 240pp. 5⅜ x 8½.
28690-8

Engineering

DE RE METALLICA, Georgius Agricola. The famous Hoover translation of greatest treatise on technological chemistry, engineering, geology, mining of early modern times (1556). All 289 original woodcuts. 638pp. 6¾ x 11.   60006-8

FUNDAMENTALS OF ASTRODYNAMICS, Roger Bate et al. Modern approach developed by U.S. Air Force Academy. Designed as a first course. Problems, exercises. Numerous illustrations. 455pp. 5⅜ x 8½.   60061-0

DYNAMICS OF FLUIDS IN POROUS MEDIA, Jacob Bear. For advanced students of ground water hydrology, soil mechanics and physics, drainage and irrigation engineering, and more. 335 illustrations. Exercises, with answers. 784pp. 6⅛ x 9¼.
            65675-6

THEORY OF VISCOELASTICITY (Second Edition), Richard M. Christensen. Complete, consistent description of the linear theory of the viscoelastic behavior of materials. Problem-solving techniques discussed. 1982 edition. 29 figures. xiv+364pp. 6⅛ x 9¼.   42880-X

MECHANICS, J. P. Den Hartog. A classic introductory text or refresher. Hundreds of applications and design problems illuminate fundamentals of trusses, loaded beams and cables, etc. 334 answered problems. 462pp. 5⅜ x 8½.   60754-2

MECHANICAL VIBRATIONS, J. P. Den Hartog. Classic textbook offers lucid explanations and illustrative models, applying theories of vibrations to a variety of practical industrial engineering problems. Numerous figures. 233 problems, solutions. Appendix. Index. Preface. 436pp. 5⅜ x 8½.   64785-4

STRENGTH OF MATERIALS, J. P. Den Hartog. Full, clear treatment of basic material (tension, torsion, bending, etc.) plus advanced material on engineering methods, applications. 350 answered problems. 323pp. 5⅜ x 8½.   60755-0

A HISTORY OF MECHANICS, René Dugas. Monumental study of mechanical principles from antiquity to quantum mechanics. Contributions of ancient Greeks, Galileo, Leonardo, Kepler, Lagrange, many others. 671pp. 5⅜ x 8½.   65632-2

STABILITY THEORY AND ITS APPLICATIONS TO STRUCTURAL MECHANICS, Clive L. Dym. Self-contained text focuses on Koiter postbuckling analyses, with mathematical notions of stability of motion. Basing minimum energy principles for static stability upon dynamic concepts of stability of motion, it develops asymptotic buckling and postbuckling analyses from potential energy considerations, with applications to columns, plates, and arches. 1974 ed. 208pp. 5⅜ x 8½.
            42541-X

METAL FATIGUE, N. E. Frost, K. J. Marsh, and L. P. Pook. Definitive, clearly written, and well-illustrated volume addresses all aspects of the subject, from the historical development of understanding metal fatigue to vital concepts of the cyclic stress that causes a crack to grow. Includes 7 appendixes. 544pp. 5⅜ x 8½.   40927-9

ROCKETS, Robert Goddard. Two of the most significant publications in the history of rocketry and jet propulsion: "A Method of Reaching Extreme Altitudes" (1919) and "Liquid Propellant Rocket Development" (1936). 128pp. 5⅜ x 8½.   42537-1

STATISTICAL MECHANICS: Principles and Applications, Terrell L. Hill. Standard text covers fundamentals of statistical mechanics, applications to fluctuation theory, imperfect gases, distribution functions, more. 448pp. 5⅜ x 8½.   65390-0

ENGINEERING AND TECHNOLOGY 1650–1750: Illustrations and Texts from Original Sources, Martin Jensen. Highly readable text with more than 200 contemporary drawings and detailed engravings of engineering projects dealing with surveying, leveling, materials, hand tools, lifting equipment, transport and erection, piling, bailing, water supply, hydraulic engineering, and more. Among the specific projects outlined–transporting a 50-ton stone to the Louvre, erecting an obelisk, building timber locks, and dredging canals. 207pp. 8⅜ x 11¼.   42232-1

THE VARIATIONAL PRINCIPLES OF MECHANICS, Cornelius Lanczos. Graduate level coverage of calculus of variations, equations of motion, relativistic mechanics, more. First inexpensive paperbound edition of classic treatise. Index. Bibliography. 418pp. 5⅜ x 8½.   65067-7

PROTECTION OF ELECTRONIC CIRCUITS FROM OVERVOLTAGES, Ronald B. Standler. Five-part treatment presents practical rules and strategies for circuits designed to protect electronic systems from damage by transient overvoltages. 1989 ed. xxiv+434pp. 6⅛ x 9¼.   42552-5

ROTARY WING AERODYNAMICS, W. Z. Stepniewski. Clear, concise text covers aerodynamic phenomena of the rotor and offers guidelines for helicopter performance evaluation. Originally prepared for NASA. 537 figures. 640pp. 6⅛ x 9¼.
        64647-5

INTRODUCTION TO SPACE DYNAMICS, William Tyrrell Thomson. Comprehensive, classic introduction to space-flight engineering for advanced undergraduate and graduate students. Includes vector algebra, kinematics, transformation of coordinates. Bibliography. Index. 352pp. 5⅜ x 8½.   65113-4

HISTORY OF STRENGTH OF MATERIALS, Stephen P. Timoshenko. Excellent historical survey of the strength of materials with many references to the theories of elasticity and structure. 245 figures. 452pp. 5⅜ x 8½.   61187-6

ANALYTICAL FRACTURE MECHANICS, David J. Unger. Self-contained text supplements standard fracture mechanics texts by focusing on analytical methods for determining crack-tip stress and strain fields. 336pp. 6⅛ x 9¼.   41737-9

STATISTICAL MECHANICS OF ELASTICITY, J. H. Weiner. Advanced, self-contained treatment illustrates general principles and elastic behavior of solids. Part 1, based on classical mechanics, studies thermoelastic behavior of crystalline and polymeric solids. Part 2, based on quantum mechanics, focuses on interatomic force laws, behavior of solids, and thermally activated processes. For students of physics and chemistry and for polymer physicists. 1983 ed. 96 figures. 496pp. 5⅜ x 8½.   42260-7

Mathematics

FUNCTIONAL ANALYSIS (Second Corrected Edition), George Bachman and Lawrence Narici. Excellent treatment of subject geared toward students with background in linear algebra, advanced calculus, physics, and engineering. Text covers introduction to inner-product spaces, normed, metric spaces, and topological spaces; complete orthonormal sets, the Hahn-Banach Theorem and its consequences, and many other related subjects. 1966 ed. 544pp. 6⅛ x 9¼. 40251-7

ASYMPTOTIC EXPANSIONS OF INTEGRALS, Norman Bleistein & Richard A. Handelsman. Best introduction to important field with applications in a variety of scientific disciplines. New preface. Problems. Diagrams. Tables. Bibliography. Index. 448pp. 5⅜ x 8½. 65082-0

VECTOR AND TENSOR ANALYSIS WITH APPLICATIONS, A. I. Borisenko and I. E. Tarapov. Concise introduction. Worked-out problems, solutions, exercises. 257pp. 5⅝ x 8¼. 63833-2

THE ABSOLUTE DIFFERENTIAL CALCULUS (CALCULUS OF TENSORS), Tullio Levi-Civita. Great 20th-century mathematician's classic work on material necessary for mathematical grasp of theory of relativity. 452pp. 5⅝ x 8¼. 63401-9

AN INTRODUCTION TO ORDINARY DIFFERENTIAL EQUATIONS, Earl A. Coddington. A thorough and systematic first course in elementary differential equations for undergraduates in mathematics and science, with many exercises and problems (with answers). Index. 304pp. 5⅜ x 8½. 65942-9

FOURIER SERIES AND ORTHOGONAL FUNCTIONS, Harry F. Davis. An incisive text combining theory and practical example to introduce Fourier series, orthogonal functions and applications of the Fourier method to boundary-value problems. 570 exercises. Answers and notes. 416pp. 5⅜ x 8½. 65973-9

COMPUTABILITY AND UNSOLVABILITY, Martin Davis. Classic graduate-level introduction to theory of computability, usually referred to as theory of recurrent functions. New preface and appendix. 288pp. 5⅜ x 8½. 61471-9

ASYMPTOTIC METHODS IN ANALYSIS, N. G. de Bruijn. An inexpensive, comprehensive guide to asymptotic methods–the pioneering work that teaches by explaining worked examples in detail. Index. 224pp. 5⅜ x 8½ 64221-6

APPLIED COMPLEX VARIABLES, John W. Dettman. Step-by-step coverage of fundamentals of analytic function theory–plus lucid exposition of five important applications: Potential Theory; Ordinary Differential Equations; Fourier Transforms; Laplace Transforms; Asymptotic Expansions. 66 figures. Exercises at chapter ends. 512pp. 5⅜ x 8½. 64670-X

INTRODUCTION TO LINEAR ALGEBRA AND DIFFERENTIAL EQUATIONS, John W. Dettman. Excellent text covers complex numbers, determinants, orthonormal bases, Laplace transforms, much more. Exercises with solutions. Undergraduate level. 416pp. 5⅜ x 8½. 65191-6

CALCULUS OF VARIATIONS WITH APPLICATIONS, George M. Ewing. Applications-oriented introduction to variational theory develops insight and promotes understanding of specialized books, research papers. Suitable for advanced undergraduate/graduate students as primary, supplementary text. 352pp. 5⅜ x 8½.
64856-7

COMPLEX VARIABLES, Francis J. Flanigan. Unusual approach, delaying complex algebra till harmonic functions have been analyzed from real variable viewpoint. Includes problems with answers. 364pp. 5⅜ x 8½.
61388-7

AN INTRODUCTION TO THE CALCULUS OF VARIATIONS, Charles Fox. Graduate-level text covers variations of an integral, isoperimetrical problems, least action, special relativity, approximations, more. References. 279pp. 5⅜ x 8½.
65499-0

COUNTEREXAMPLES IN ANALYSIS, Bernard R. Gelbaum and John M. H. Olmsted. These counterexamples deal mostly with the part of analysis known as "real variables." The first half covers the real number system, and the second half encompasses higher dimensions. 1962 edition. xxiv+198pp. 5⅜ x 8½.
42875-3

CATASTROPHE THEORY FOR SCIENTISTS AND ENGINEERS, Robert Gilmore. Advanced-level treatment describes mathematics of theory grounded in the work of Poincaré, R. Thom, other mathematicians. Also important applications to problems in mathematics, physics, chemistry, and engineering. 1981 edition. References. 28 tables. 397 black-and-white illustrations. xvii+666pp. 6⅛ x 9¼.
67539-4

INTRODUCTION TO DIFFERENCE EQUATIONS, Samuel Goldberg. Exceptionally clear exposition of important discipline with applications to sociology, psychology, economics. Many illustrative examples; over 250 problems. 260pp. 5⅜ x 8½.
65084-7

NUMERICAL METHODS FOR SCIENTISTS AND ENGINEERS, Richard Hamming. Classic text stresses frequency approach in coverage of algorithms, polynomial approximation, Fourier approximation, exponential approximation, other topics. Revised and enlarged 2nd edition. 721pp. 5⅜ x 8½.
65241-6

INTRODUCTION TO NUMERICAL ANALYSIS (2nd Edition), F. B. Hildebrand. Classic, fundamental treatment covers computation, approximation, interpolation, numerical differentiation and integration, other topics. 150 new problems. 669pp. 5⅜ x 8½.
65363-3

THREE PEARLS OF NUMBER THEORY, A. Y. Khinchin. Three compelling puzzles require proof of a basic law governing the world of numbers. Challenges concern van der Waerden's theorem, the Landau-Schnirelmann hypothesis and Mann's theorem, and a solution to Waring's problem. Solutions included. 64pp. 5⅜ x 8½.
40026-3

THE PHILOSOPHY OF MATHEMATICS: An Introductory Essay, Stephan Körner. Surveys the views of Plato, Aristotle, Leibniz & Kant concerning propositions and theories of applied and pure mathematics. Introduction. Two appendices. Index. 198pp. 5⅜ x 8½.
25048-2

CATALOG OF DOVER BOOKS

INTRODUCTORY REAL ANALYSIS, A.N. Kolmogorov, S. V. Fomin. Translated by Richard A. Silverman. Self-contained, evenly paced introduction to real and functional analysis. Some 350 problems. 403pp. 5⅜ x 8½. 61226-0

APPLIED ANALYSIS, Cornelius Lanczos. Classic work on analysis and design of finite processes for approximating solution of analytical problems. Algebraic equations, matrices, harmonic analysis, quadrature methods, more. 559pp. 5⅜ x 8½. 65656-X

AN INTRODUCTION TO ALGEBRAIC STRUCTURES, Joseph Landin. Superb self-contained text covers "abstract algebra": sets and numbers, theory of groups, theory of rings, much more. Numerous well-chosen examples, exercises. 247pp. 5⅜ x 8½. 65940-2

QUALITATIVE THEORY OF DIFFERENTIAL EQUATIONS, V. V. Nemytskii and V.V. Stepanov. Classic graduate-level text by two prominent Soviet mathematicians covers classical differential equations as well as topological dynamics and ergodic theory. Bibliographies. 523pp. 5⅜ x 8½. 65954-2

THEORY OF MATRICES, Sam Perlis. Outstanding text covering rank, nonsingularity and inverses in connection with the development of canonical matrices under the relation of equivalence, and without the intervention of determinants. Includes exercises. 237pp. 5⅜ x 8½. 66810-X

INTRODUCTION TO ANALYSIS, Maxwell Rosenlicht. Unusually clear, accessible coverage of set theory, real number system, metric spaces, continuous functions, Riemann integration, multiple integrals, more. Wide range of problems. Undergraduate level. Bibliography. 254pp. 5⅜ x 8½. 65038-3

MODERN NONLINEAR EQUATIONS, Thomas L. Saaty. Emphasizes practical solution of problems; covers seven types of equations. ". . . a welcome contribution to the existing literature. . . ."–Math Reviews. 490pp. 5⅜ x 8½. 64232-1

MATRICES AND LINEAR ALGEBRA, Hans Schneider and George Phillip Barker. Basic textbook covers theory of matrices and its applications to systems of linear equations and related topics such as determinants, eigenvalues, and differential equations. Numerous exercises. 432pp. 5⅜ x 8½. 66014-1

MATHEMATICS APPLIED TO CONTINUUM MECHANICS, Lee A. Segel. Analyzes models of fluid flow and solid deformation. For upper-level math, science, and engineering students. 608pp. 5⅜ x 8½. 65369-2

ELEMENTS OF REAL ANALYSIS, David A. Sprecher. Classic text covers fundamental concepts, real number system, point sets, functions of a real variable, Fourier series, much more. Over 500 exercises. 352pp. 5⅜ x 8½. 65385-4

SET THEORY AND LOGIC, Robert R. Stoll. Lucid introduction to unified theory of mathematical concepts. Set theory and logic seen as tools for conceptual understanding of real number system. 496pp. 5⅜ x 8¼. 63829-4

TENSOR CALCULUS, J.L. Synge and A. Schild. Widely used introductory text covers spaces and tensors, basic operations in Riemannian space, non-Riemannian spaces, etc. 324pp. 5⅝ x 8¼. 63612-7

ORDINARY DIFFERENTIAL EQUATIONS, Morris Tenenbaum and Harry Pollard. Exhaustive survey of ordinary differential equations for undergraduates in mathematics, engineering, science. Thorough analysis of theorems. Diagrams. Bibliography. Index. 818pp. 5⅜ x 8½. 64940-7

INTEGRAL EQUATIONS, F. G. Tricomi. Authoritative, well-written treatment of extremely useful mathematical tool with wide applications. Volterra Equations, Fredholm Equations, much more. Advanced undergraduate to graduate level. Exercises. Bibliography. 238pp. 5⅜ x 8½. 64828-1

FOURIER SERIES, Georgi P. Tolstov. Translated by Richard A. Silverman. A valuable addition to the literature on the subject, moving clearly from subject to subject and theorem to theorem. 107 problems, answers. 336pp. 5⅜ x 8½. 63317-9

INTRODUCTION TO MATHEMATICAL THINKING, Friedrich Waismann. Examinations of arithmetic, geometry, and theory of integers; rational and natural numbers; complete induction; limit and point of accumulation; remarkable curves; complex and hypercomplex numbers, more. 1959 ed. 27 figures. xii+260pp. 5⅜ x 8½. 42804-4

POPULAR LECTURES ON MATHEMATICAL LOGIC, Hao Wang. Noted logician's lucid treatment of historical developments, set theory, model theory, recursion theory and constructivism, proof theory, more. 3 appendixes. Bibliography. 1981 ed. ix+283pp. 5⅜ x 8½. 67632-3

CALCULUS OF VARIATIONS, Robert Weinstock. Basic introduction covering isoperimetric problems, theory of elasticity, quantum mechanics, electrostatics, etc. Exercises throughout. 326pp. 5⅜ x 8½. 63069-2

THE CONTINUUM: A Critical Examination of the Foundation of Analysis, Hermann Weyl. Classic of 20th-century foundational research deals with the conceptual problem posed by the continuum. 156pp. 5⅜ x 8½. 67982-9

CHALLENGING MATHEMATICAL PROBLEMS WITH ELEMENTARY SOLUTIONS, A. M. Yaglom and I. M. Yaglom. Over 170 challenging problems on probability theory, combinatorial analysis, points and lines, topology, convex polygons, many other topics. Solutions. Total of 445pp. 5⅜ x 8½. Two-vol. set.
Vol. I: 65536-9 Vol. II: 65537-7

INTRODUCTION TO PARTIAL DIFFERENTIAL EQUATIONS WITH APPLICATIONS, E. C. Zachmanoglou and Dale W. Thoe. Essentials of partial differential equations applied to common problems in engineering and the physical sciences. Problems and answers. 416pp. 5⅜ x 8½. 65251-3

THE THEORY OF GROUPS, Hans J. Zassenhaus. Well-written graduate-level text acquaints reader with group-theoretic methods and demonstrates their usefulness in mathematics. Axioms, the calculus of complexes, homomorphic mapping, *p*-group theory, more. 276pp. 5⅜ x 8½. 40922-8

Math–Decision Theory, Statistics, Probability

ELEMENTARY DECISION THEORY, Herman Chernoff and Lincoln E. Moses. Clear introduction to statistics and statistical theory covers data processing, probability and random variables, testing hypotheses, much more. Exercises. 364pp. 5⅜ x 8½. 65218-1

STATISTICS MANUAL, Edwin L. Crow et al. Comprehensive, practical collection of classical and modern methods prepared by U.S. Naval Ordnance Test Station. Stress on use. Basics of statistics assumed. 288pp. 5⅜ x 8½. 60599-X

SOME THEORY OF SAMPLING, William Edwards Deming. Analysis of the problems, theory, and design of sampling techniques for social scientists, industrial managers, and others who find statistics important at work. 61 tables. 90 figures. xvii +602pp. 5⅜ x 8½. 64684-X

LINEAR PROGRAMMING AND ECONOMIC ANALYSIS, Robert Dorfman, Paul A. Samuelson and Robert M. Solow. First comprehensive treatment of linear programming in standard economic analysis. Game theory, modern welfare economics, Leontief input-output, more. 525pp. 5⅜ x 8½. 65491-5

PROBABILITY: An Introduction, Samuel Goldberg. Excellent basic text covers set theory, probability theory for finite sample spaces, binomial theorem, much more. 360 problems. Bibliographies. 322pp. 5⅜ x 8½. 65252-1

GAMES AND DECISIONS: Introduction and Critical Survey, R. Duncan Luce and Howard Raiffa. Superb nontechnical introduction to game theory, primarily applied to social sciences. Utility theory, zero-sum games, n-person games, decision-making, much more. Bibliography. 509pp. 5⅜ x 8½. 65943-7

INTRODUCTION TO THE THEORY OF GAMES, J. C. C. McKinsey. This comprehensive overview of the mathematical theory of games illustrates applications to situations involving conflicts of interest, including economic, social, political, and military contexts. Appropriate for advanced undergraduate and graduate courses; advanced calculus a prerequisite. 1952 ed. x+372pp. 5⅜ x 8½. 42811-7

FIFTY CHALLENGING PROBLEMS IN PROBABILITY WITH SOLUTIONS, Frederick Mosteller. Remarkable puzzlers, graded in difficulty, illustrate elementary and advanced aspects of probability. Detailed solutions. 88pp. 5⅜ x 8½. 65355-2

PROBABILITY THEORY: A Concise Course, Y. A. Rozanov. Highly readable, self-contained introduction covers combination of events, dependent events, Bernoulli trials, etc. 148pp. 5⅜ x 8¼. 63544-9

STATISTICAL METHOD FROM THE VIEWPOINT OF QUALITY CONTROL, Walter A. Shewhart. Important text explains regulation of variables, uses of statistical control to achieve quality control in industry, agriculture, other areas. 192pp. 5⅜ x 8½. 65232-7

Math–Geometry and Topology

ELEMENTARY CONCEPTS OF TOPOLOGY, Paul Alexandroff. Elegant, intuitive approach to topology from set-theoretic topology to Betti groups; how concepts of topology are useful in math and physics. 25 figures. 57pp. 5⅜ x 8½. 60747-X

COMBINATORIAL TOPOLOGY, P. S. Alexandrov. Clearly written, well-organized, three-part text begins by dealing with certain classic problems without using the formal techniques of homology theory and advances to the central concept, the Betti groups. Numerous detailed examples. 654pp. 5⅜ x 8½. 40179-0

EXPERIMENTS IN TOPOLOGY, Stephen Barr. Classic, lively explanation of one of the byways of mathematics. Klein bottles, Moebius strips, projective planes, map coloring, problem of the Koenigsberg bridges, much more, described with clarity and wit. 43 figures. 210pp. 5⅜ x 8½. 25933-1

CONFORMAL MAPPING ON RIEMANN SURFACES, Harvey Cohn. Lucid, insightful book presents ideal coverage of subject. 334 exercises make book perfect for self-study. 55 figures. 352pp. 5⅝ x 8¼. 64025-6

THE GEOMETRY OF RENÉ DESCARTES, René Descartes. The great work founded analytical geometry. Original French text, Descartes's own diagrams, together with definitive Smith-Latham translation. 244pp. 5⅜ x 8½. 60068-8

PRACTICAL CONIC SECTIONS: The Geometric Properties of Ellipses, Parabolas and Hyperbolas, J. W. Downs. This text shows how to create ellipses, parabolas, and hyperbolas. It also presents historical background on their ancient origins and describes the reflective properties and roles of curves in design applications. 1993 ed. 98 figures. xii+100pp. 6½ x 9¼. 42876-1

THE THIRTEEN BOOKS OF EUCLID'S ELEMENTS, translated with introduction and commentary by Thomas L. Heath. Definitive edition. Textual and linguistic notes, mathematical analysis. 2,500 years of critical commentary. Unabridged. 1,414pp. 5⅜ x 8½. Three-vol. set. Vol. I: 60088-2 Vol. II: 60089-0 Vol. III: 60090-4

GEOMETRY OF COMPLEX NUMBERS, Hans Schwerdtfeger. Illuminating, widely praised book on analytic geometry of circles, the Moebius transformation, and two-dimensional non-Euclidean geometries. 200pp. 5⅝ x 8¼. 63830-8

DIFFERENTIAL GEOMETRY, Heinrich W. Guggenheimer. Local differential geometry as an application of advanced calculus and linear algebra. Curvature, transformation groups, surfaces, more. Exercises. 62 figures. 378pp. 5⅜ x 8½. 63433-7

CURVATURE AND HOMOLOGY: Enlarged Edition, Samuel I. Goldberg. Revised edition examines topology of differentiable manifolds; curvature, homology of Riemannian manifolds; compact Lie groups; complex manifolds; curvature, homology of Kaehler manifolds. New Preface. Four new appendixes. 416pp. 5⅜ x 8½. 40207-X

History of Math

THE WORKS OF ARCHIMEDES, Archimedes (T. L. Heath, ed.). Topics include the famous problems of the ratio of the areas of a cylinder and an inscribed sphere; the measurement of a circle; the properties of conoids, spheroids, and spirals; and the quadrature of the parabola. Informative introduction. clxxxvi+326pp; supplement, 52pp. 5⅜ x 8½.   42084-1

A SHORT ACCOUNT OF THE HISTORY OF MATHEMATICS, W. W. Rouse Ball. One of clearest, most authoritative surveys from the Egyptians and Phoenicians through 19th-century figures such as Grassman, Galois, Riemann. Fourth edition. 522pp. 5⅜ x 8½.   20630-0

THE HISTORY OF THE CALCULUS AND ITS CONCEPTUAL DEVELOP-MENT, Carl B. Boyer. Origins in antiquity, medieval contributions, work of Newton, Leibniz, rigorous formulation. Treatment is verbal. 346pp. 5⅜ x 8½.   60509-4

THE HISTORICAL ROOTS OF ELEMENTARY MATHEMATICS, Lucas N. H. Bunt, Phillip S. Jones, and Jack D. Bedient. Fundamental underpinnings of modern arithmetic, algebra, geometry, and number systems derived from ancient civiliza-tions. 320pp. 5⅜ x 8½.   25563-8

A HISTORY OF MATHEMATICAL NOTATIONS, Florian Cajori. This classic study notes the first appearance of a mathematical symbol and its origin, the com-petition it encountered, its spread among writers in different countries, its rise to pop-ularity, its eventual decline or ultimate survival. Original 1929 two-volume edition presented here in one volume. xxviii+820pp. 5⅜ x 8½.   67766-4

GAMES, GODS & GAMBLING: A History of Probability and Statistical Ideas, F. N. David. Episodes from the lives of Galileo, Fermat, Pascal, and others illustrate this fascinating account of the roots of mathematics. Features thought-provoking refer-ences to classics, archaeology, biography, poetry. 1962 edition. 304pp. 5⅜ x 8½. (Available in U.S. only.)   40023-9

OF MEN AND NUMBERS: The Story of the Great Mathematicians, Jane Muir. Fascinating accounts of the lives and accomplishments of history's greatest mathe-matical minds–Pythagoras, Descartes, Euler, Pascal, Cantor, many more. Anecdotal, illuminating. 30 diagrams. Bibliography. 256pp. 5⅜ x 8½.   28973-7

HISTORY OF MATHEMATICS, David E. Smith. Nontechnical survey from ancient Greece and Orient to late 19th century; evolution of arithmetic, geometry, trigonometry, calculating devices, algebra, the calculus. 362 illustrations. 1,355pp. 5⅜ x 8½. Two-vol. set.   Vol. I: 20429-4   Vol. II: 20430-8

A CONCISE HISTORY OF MATHEMATICS, Dirk J. Struik. The best brief his-tory of mathematics. Stresses origins and covers every major figure from ancient Near East to 19th century. 41 illustrations. 195pp. 5⅜ x 8½.   60255-9

Physics

OPTICAL RESONANCE AND TWO-LEVEL ATOMS, L. Allen and J. H. Eberly. Clear, comprehensive introduction to basic principles behind all quantum optical resonance phenomena. 53 illustrations. Preface. Index. 256pp. 5⅜ x 8½. 65533-4

QUANTUM THEORY, David Bohm. This advanced undergraduate-level text presents the quantum theory in terms of qualitative and imaginative concepts, followed by specific applications worked out in mathematical detail. Preface. Index. 655pp. 5⅜ x 8½. 65969-0

ATOMIC PHYSICS: 8th edition, Max Born. Nobel laureate's lucid treatment of kinetic theory of gases, elementary particles, nuclear atom, wave-corpuscles, atomic structure and spectral lines, much more. Over 40 appendices, bibliography. 495pp. 5⅜ x 8½. 65984-4

A SOPHISTICATE'S PRIMER OF RELATIVITY, P. W. Bridgman. Geared toward readers already acquainted with special relativity, this book transcends the view of theory as a working tool to answer natural questions: What is a frame of reference? What is a "law of nature"? What is the role of the "observer"? Extensive treatment, written in terms accessible to those without a scientific background. 1983 ed. xlviii+172pp. 5⅜ x 8½. 42549-5

AN INTRODUCTION TO HAMILTONIAN OPTICS, H. A. Buchdahl. Detailed account of the Hamiltonian treatment of aberration theory in geometrical optics. Many classes of optical systems defined in terms of the symmetries they possess. Problems with detailed solutions. 1970 edition. xv+360pp. 5⅜ x 8½. 67597-1

PRIMER OF QUANTUM MECHANICS, Marvin Chester. Introductory text examines the classical quantum bead on a track: its state and representations; operator eigenvalues; harmonic oscillator and bound bead in a symmetric force field; and bead in a spherical shell. Other topics include spin, matrices, and the structure of quantum mechanics; the simplest atom; indistinguishable particles; and stationary-state perturbation theory. 1992 ed. xiv+314pp. 6⅛ x 9¼. 42878-8

LECTURES ON QUANTUM MECHANICS, Paul A. M. Dirac. Four concise, brilliant lectures on mathematical methods in quantum mechanics from Nobel Prize–winning quantum pioneer build on idea of visualizing quantum theory through the use of classical mechanics. 96pp. 5⅜ x 8½. 41713-1

THIRTY YEARS THAT SHOOK PHYSICS: The Story of Quantum Theory, George Gamow. Lucid, accessible introduction to influential theory of energy and matter. Careful explanations of Dirac's anti-particles, Bohr's model of the atom, much more. 12 plates. Numerous drawings. 240pp. 5⅜ x 8½. 24895-X

ELECTRONIC STRUCTURE AND THE PROPERTIES OF SOLIDS: The Physics of the Chemical Bond, Walter A. Harrison. Innovative text offers basic understanding of the electronic structure of covalent and ionic solids, simple metals, transition metals and their compounds. Problems. 1980 edition. 582pp. 6⅛ x 9¼. 66021-4

CATALOG OF DOVER BOOKS

HYDRODYNAMIC AND HYDROMAGNETIC STABILITY, S. Chandrasekhar. Lucid examination of the Rayleigh-Benard problem; clear coverage of the theory of instabilities causing convection. 704pp. 5⅜ x 8¼. 64071-X

INVESTIGATIONS ON THE THEORY OF THE BROWNIAN MOVEMENT, Albert Einstein. Five papers (1905–8) investigating dynamics of Brownian motion and evolving elementary theory. Notes by R. Fürth. 122pp. 5⅜ x 8½. 60304-0

THE PHYSICS OF WAVES, William C. Elmore and Mark A. Heald. Unique overview of classical wave theory. Acoustics, optics, electromagnetic radiation, more. Ideal as classroom text or for self-study. Problems. 477pp. 5⅜ x 8½. 64926-1

PHYSICAL PRINCIPLES OF THE QUANTUM THEORY, Werner Heisenberg. Nobel Laureate discusses quantum theory, uncertainty, wave mechanics, work of Dirac, Schroedinger, Compton, Wilson, Einstein, etc. 184pp. 5⅜ x 8½. 60113-7

ATOMIC SPECTRA AND ATOMIC STRUCTURE, Gerhard Herzberg. One of best introductions; especially for specialist in other fields. Treatment is physical rather than mathematical. 80 illustrations. 257pp. 5⅜ x 8½. 60115-3

AN INTRODUCTION TO STATISTICAL THERMODYNAMICS, Terrell L. Hill. Excellent basic text offers wide-ranging coverage of quantum statistical mechanics, systems of interacting molecules, quantum statistics, more. 523pp. 5⅜ x 8½. 65242-4

THEORETICAL PHYSICS, Georg Joos, with Ira M. Freeman. Classic overview covers essential math, mechanics, electromagnetic theory, thermodynamics, quantum mechanics, nuclear physics, other topics. xxiii+885pp. 5⅜ x 8½. 65227-0

PROBLEMS AND SOLUTIONS IN QUANTUM CHEMISTRY AND PHYSICS, Charles S. Johnson, Jr. and Lee G. Pedersen. Unusually varied problems, detailed solutions in coverage of quantum mechanics, wave mechanics, angular momentum, molecular spectroscopy, more. 280 problems, 139 supplementary exercises. 430pp. 6½ x 9¼. 65236-X

THEORETICAL SOLID STATE PHYSICS, Vol. I: Perfect Lattices in Equilibrium; Vol. II: Non-Equilibrium and Disorder, William Jones and Norman H. March. Monumental reference work covers fundamental theory of equilibrium properties of perfect crystalline solids, non-equilibrium properties, defects and disordered systems. Total of 1,301pp. 5⅜ x 8½. Vol. I: 65015-4 Vol. II: 65016-2

WHAT IS RELATIVITY? L. D. Landau and G. B. Rumer. Written by a Nobel Prize physicist and his distinguished colleague, this compelling book explains the special theory of relativity to readers with no scientific background, using such familiar objects as trains, rulers, and clocks. 1960 ed. vi+72pp. 23 b/w illustrations. 5⅜ x 8½. 42806-0 $6.95

A TREATISE ON ELECTRICITY AND MAGNETISM, James Clerk Maxwell. Important foundation work of modern physics. Brings to final form Maxwell's theory of electromagnetism and rigorously derives his general equations of field theory. 1,084pp. 5⅜ x 8½. Two-vol. set. Vol. I: 60636-8 Vol. II: 60637-6

CATALOG OF DOVER BOOKS

QUANTUM MECHANICS: Principles and Formalism, Roy McWeeny. Graduate student–oriented volume develops subject as fundamental discipline, opening with review of origins of Schrödinger's equations and vector spaces. Focusing on main principles of quantum mechanics and their immediate consequences, it concludes with final generalizations covering alternative "languages" or representations. 1972 ed. 15 figures. xi+155pp. 5⅜ x 8½.   42829-X

INTRODUCTION TO QUANTUM MECHANICS WITH APPLICATIONS TO CHEMISTRY, Linus Pauling & E. Bright Wilson, Jr. Classic undergraduate text by Nobel Prize winner applies quantum mechanics to chemical and physical problems. Numerous tables and figures enhance the text. Chapter bibliographies. Appendices. Index. 468pp. 5⅜ x 8½.   64871-0

METHODS OF THERMODYNAMICS, Howard Reiss. Outstanding text focuses on physical technique of thermodynamics, typical problem areas of understanding, and significance and use of thermodynamic potential. 1965 edition. 238pp. 5⅜ x 8½.   69445-3

TENSOR ANALYSIS FOR PHYSICISTS, J. A. Schouten. Concise exposition of the mathematical basis of tensor analysis, integrated with well-chosen physical examples of the theory. Exercises. Index. Bibliography. 289pp. 5⅜ x 8½.   65582-2

THE ELECTROMAGNETIC FIELD, Albert Shadowitz. Comprehensive undergraduate text covers basics of electric and magnetic fields, builds up to electromagnetic theory. Also related topics, including relativity. Over 900 problems. 768pp. 5⅜ x 8¼.   65660-8

GREAT EXPERIMENTS IN PHYSICS: Firsthand Accounts from Galileo to Einstein, Morris H. Shamos (ed.). 25 crucial discoveries: Newton's laws of motion, Chadwick's study of the neutron, Hertz on electromagnetic waves, more. Original accounts clearly annotated. 370pp. 5⅜ x 8½.   25346-5

RELATIVITY, THERMODYNAMICS AND COSMOLOGY, Richard C. Tolman. Landmark study extends thermodynamics to special, general relativity; also applications of relativistic mechanics, thermodynamics to cosmological models. 501pp. 5⅜ x 8½.   65383-8

STATISTICAL PHYSICS, Gregory H. Wannier. Classic text combines thermodynamics, statistical mechanics, and kinetic theory in one unified presentation of thermal physics. Problems with solutions. Bibliography. 532pp. 5⅜ x 8½.   65401-X